History of Building

History of Building
Second Edition

JACK BOWYER
Dipl.Arch. (Leeds)

ATTIC

BOOKS

First published in 1973 by Granada Publishing Ltd.
Second edition published in 1983 by Orion Books, Eastbourne.
Reprinted in 1993 by Attic Books,
The Folly, Rhosgoch, Builth Wells, Powys LD2 3JY

British Library Cataloguing in Publication Data

   Bowyer, Jack 1927–
   History of building – (New ed.)
   1. Europe. Building. Technology
   I. Title
ISBN 0 948083 18 2 paper

Printed in Great Britain by J. W. Arrowsmith Ltd., Bristol BS3 2NT

Contents

Figures 103, 105, 106, 107, 109, 110, 111, 112, 132, 133, 138 and 139 are reproduced from *Der Vollkommen Architectus* by Andreas Grote, by kind permission of the publishers, Prestel Verlag, Munich.

Introduction

The historical development of structural form is one of slow growth, of experiment, success and failure. Occasionally, a unique achievement such as the construction of the Hagia Sophia in Constantinople proclaims the limit of a particular structural development. Usually the limitations of available knowledge and material restrict this natural evolution. Within the last hundred years steel, reinforced concrete and aluminium have brought about a revolution in structural design, not only through the properties of the materials themselves, but also by an increasing knowledge and experience in the application of the principles of structural design.

The basic needs of man are food and clothing, and shelter from the heat of the sun, from rain and cold. After the utilisation of natural rock shelters, which in some areas of the world continues to this day, man's first attempts to provide shelter were simple and crude: screens and huts of branches and twigs, walls of turf or stone slabs. He acquired, over a long period of time and by trial and error, a practical knowledge of the principles of building, the stabilisation of structure. During this period man continued to explore and develop new and better systems of construction and materials with which to build and decorate his buildings.

The basic function of a building is to provide a supporting structure to enable loads to be supported over the sheltered space and to transmit these loads with safety and economy to the ground. The designers first decision is, therefore, the selection of the type of structure to be used. This affects not only the functioning of the plan but also the value of the building. In historical studies it will be found that the development of structure has always been the point of departure from which buildings have attempted to meet the social needs of the age. Up to the end of the 18th cent., always excepting Roman concrete construction, all buildings relied on masonry or timber for their supporting structure. The principles involved were very simple, either used singly as in Greek post and lintol construction, or combining one or more which can best be seen in Roman composite work where classical post and lintol incorporated with advantage the masonry relieving arch.

Fig. 1. Monolith   Fig. 2. Dolmen

Loads imposed on and provided by a structure set up forces which must be balanced or resolved to enable the building to retain its equilibrium. In traditional buildings, all the necessary decisions to achieve this were taken by the craftsmen or, in later days, by the architect. Now that structures have reached a high degree of complexity and sophistication these decisions are made by the structural engineer specialising in this field of professional activity. It is not known exactly when the calculation of forces began to replace empirical methods. When thrusts are resolved within heavy masonry walls, the factor of safety is usually high. This is often uneconomic, but it should always be borne in mind that while economic factors are important, they are not entirely essential to the production of buildings of quality. While the strength of constituent materials is important, that of the joint or jointing material may well prove the weak link. The Romans made their jointing mortar of such strength that the wall and vault became a monolithic mass, a technique abandoned until the invention of modern cements in the 19th cent. During this intervening period walling units were usually of brick or stone for structures of importance. For simpler buildings, timber was widely used as a framework or supporting member, especially in domestic work. Its use here points directly to the present use of steel. Timber was nearly always used to carry the superimposed

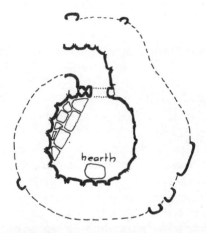

Fig. 3. Plan of a Dartmoor hut

roofing finish over stone vaulted structures as it was an admirable material to provide the necessary members to support an inclined plane. The limitations in structural knowledge, tools and jointing materials held back the logical development of timber, which only began to emerge in the 19th cent. after some interesting experiments in France with laminated construction.

Since the early years of this century, few buildings have been designed in which economy was not important. An age of experiment and inquiry has attempted to correlate economy with the life expectancy of the building, its appearance and function. The 19th cent. made adventurous use of iron and steel. Today, the use of reinforced concrete in structures of unimagined simplicity and apparent frailty provide a flexibility in use, unknown in the past. This is the field in which future advances will be most pronounced.

Whereas men like John Ruskin and William Morris studied natural forms for their beauty alone, scientific study has discovered numerous examples of efficient structural forms in nature. This is apparent in considering the egg, which uses its enveloping structure with admirable economy for the purpose it has to achieve. In nature, continuity is an important factor in this process, its emulation in building being difficult to produce. Buildings are usually composed of a conglomeration of cubes, while nature abhors a right angle. In buildings, natural forms are usually applicable to unicellular structures, the blister hanger, the parabolic reflector. In materials, the development of corrugated sheets emulates the sections of grasses. Nature creates by slow growth and evolution, man by the manipulation of materials whose inner structure is already formed (timber and natural stone), by crude fabrication by the application of heat (bricks and tiles), or by chemical reaction (concrete). The lessons to be learnt by man from natural forms are, therefore, limited to continuity of structure and the shaping of structural sections to emulate forms of proved success.

There can be no doubt but that there have been periods in the history of Western civilisation when progress in technology moved forward more rapidly than others. One such period was around 3500 BC with the evolution of, for practical purposes, the wheel and the discovery that the addition of tin to copper produced a metal capable of receiving a cutting edge. Another period was during the 5th to 3rd cents BC in Greece, when philosophers laid down many of the principles from which have originated our modern mathematics.

The Romans were not great innovators, their skill lay in the application of ideas rather than their formulation. Probably the greatest period in Western civilisation was heralded by the invention of the quadripartite vault by unknown masons at Durham Cathedral, which invention gave us the glories of Gothic architecture. All these have marked a progression in knowledge and constructional skill, a broadening of experience which forms the basis of our technology today.

As methods of fabrication changed and developed, so did the organisation necessary to put these into practice. From the soldier engineer of Roman days with his para-military building organisation based on a steady supply of enslaved labourers, to the mediaeval craft lodge supervised by the architect master mason, there continued a steady forward movement which led, through the rise of the professional designer, to the separation of building into the twin activities of design and construction, which today has become the accepted organisational division for the building industry. This has given rise to other professional activities required for the sophisticated structures considered necessary for our modern society–the structural, mechanical and electrical engineers and the measurer or quantity surveyor. In the building industry itself, changes have been taking place which, from the relatively small sub-contracting businesses of the 18th cent., have combined and amalgamated to provide the large and diversified organisations which carry out much of the constructional work today.

With organisational and structural changes have come new skills and standards. Whereas the work in Roman times was relatively unskilled, tradesmen being employed to work and fix the finishing veneers while the structural concrete was provided by the labourer, in mediaeval times the craftsman took over the bulk of the building work after the initial cutting and rough shaping by the semi-skilled stone-hewer and woodsman. Late mediaeval stone structures were a product of a system of training, technical and artistic ability only equalled by the marble masons of classical Greece. These standards, so far as carcassing trades were concerned, declined after the Renaissance, being replaced by craft skills in the decorative trades which produced the exquisite plasterwork of Georgian architecture. These skills coarsened, being replaced by mere mechanical repetition which, although satisfying the need for greater productivity, created work of little imaginative quality. The history of building is therefore a story continually unfolding to display different facets illustrating man's changing skills and needs in an ever-evolving environment.

PART ONE

Evolution of Structure

CHAPTER ONE

Post and Lintol

The earliest structures produced by man were single cell units. These provided shelter from the elements and protection from enemies and predators. Walls were of turf or wattle, and the roof was usually constructed from inclined poles thatched with furze or turf. These roofs were either self-supporting or assisted by the provision of a central pole or ridgetree.

The constructional system employing solid masonry walls pierced by small openings to provide access, air and light remained in use throughout Western civilisation. At certain periods, structural techniques were advanced which provided a new rhythm and dimension in design, a new formula for living. In many instances these techniques were monumental in character, more fitting to the formal civic structure than to the prosaic domestic. Other techniques were so universally accepted that departure from their use is worthy of comment.

Post and lintol construction is one of the earliest systems known to man. Evolving in the mists of archaic time, its use has ebbed and flowed over the centuries, never wholly out of fashion and, with few exceptions, never wholly accepted. It is probable that its cost, relative to solid masonry, has always been sufficiently high to restrict its use to structures of a civic or monumental character, and it must be admitted that it is most successful when used in this way.

The system, no doubt, evolved from timber prototypes, probably in widely separated parts of the world at about the same time. In Britain probably the earliest example is Stonehenge, consisting of a ring of upright monoliths, each providing support to a horizontal member, enclosing an inner horseshoe of similar uprights and lintols. It is more than likely that Stonehenge was copied from an earlier structure of similar size constructed from timber and discovered and excavated at Woodhenge, some two miles distant. The quality of the finish provided by the builders of Stonehenge was exceptional; incorporated in the construction were mortice and tenon joints, linking the tops of the uprights with the lintols, a feature common to classical work in the Eastern Mediterranean.

Fig. 4.   Stonehenge (after Stukeley)

The monumental structures of ancient Egypt are directly associated with timber and reed prototypes. Before the evolution of metal tools the principal building materials of Egypt, granite and limestone, were not available to the Stone Age men of the Nile valley. The materials used for early structures were reeds cut from the river and clay dug from its banks. The primitive buildings of these early Egyptians were composed of bundles of reed bound together and placed vertically in the ground at intervals. Joining these at the top were other bundles laid horizontally which bound the heads of the uprights together. The pressure on the top of the reeds caused by the weight of the clay finish to the roof resulted in the formation of the characteristic cornice. This form of construction was copied in stone during the time of the Theban Kingdom (c. 3000–2100 BC) and an example deriving from a bundle of reeds tied together at intervals and crowned by a lotus bud capital occurs in the granite columns of Thotmes III at Karnak. Although materials changed, the forms of early clay and reed construction continued to be used and it was the endeavour of the conservative Egyptians to perpetuate in stone and granite the earlier constructional appearance.

While the Egyptians employed solid masonry walling techniques, strengthened by massive pylons at the principal entrances, for the exteriors of their temples, internally they employed post and lintol construction to support the roofs of these vast structures. At the Temple of Khons at Karnak the

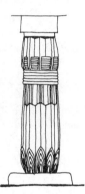

Fig. 5.   Papyrus column with lotus bud capital, Karnak (See also Fig. 10, p. 17)

entrance court, open to the sky, is surrounded on three sides by two rows of columns with lotus bud capitals supporting a web of stone lintols. These in turn support the stone slabs forming the flat roof deck. Perhaps the greatest example of Egyptian post and lintol construction is to be found in the Great Temple of Ammon, also at Karnak. Here the Hypostyle Hall measures 102×50 m (338×170 ft), the roof supported by 134 columns in 16 rows. The central avenues have columns 21 m (69 ft) high and 3·5 m (11½ ft) in diameter with lotus blossom capitals. The walls of the hall, columns and lintols (or architraves) over are covered with incised inscriptions, many of them still retaining their original coloured decoration.

Crete lies in the centre of Eastern Mediterranean cultures, but it was not until excavations were carried out at Knossos in the early years of this century by Sir Arthur Evans that the true greatness of Minoan civilisation was revealed. The constructional techniques were, in the main, similar to those of Egypt: solid walls of carefully squared masonry, and flat roofs. Large porticos and halls were supported by painted circular columns of cypress wood, tapering towards the base and provided with a boldly moulded capital and square abacus. To assist in

Fig. 6.   Cornice. Temple at Denderah, Egypt

strengthening the whole structure against earthquakes, which have been common in Crete, timber was used to form an interlacing framework, tying the masonry structure together. Although nothing like so grand and monumental as the temples of Egypt, the Minoan system of post and lintol construction with its provision for resisting earthquake shock was an important technical achievement.

The Hellenic period, which lasted from about 700 BC to the beginning of the Roman occupation in 146 BC, is notable for the development of a system of post and lintol construction known as trabeated (trabs – a beam or lintol). The style is recognised as of a special Grecian type and the character was largely influenced by the material used in its construction – finely dressed marble. Stability was achieved solely by the careful observance of gravity; all weights were designed to act vertically, no inclined thrusts were permitted. The style originated in a timber technique, and followed the effects thus produced in a carefully designed and stylised form in marble.

As in previous post and lintol techniques, marble lintols in any length were not easy to obtain and in conjunction with their low resistance to loading, necessitated relatively close spacing of the columns or upright members. It was found that by placing the stone lintols with their planes of natural cleavage in a vertical position a wider intercolumnation could be obtained. Mortar was unnecessary, as the bedding joints were rubbed to a perfectly smooth and fine surface and fixed with iron cramps. The general character of the early work was heavy and severe but in later work the proportions of the columns became more slender and the mouldings more refined. The masterpieces of Greek architecture were all erected in the period between 480 and 323 BC, when refinements in design to correct optical illusions were incorporated in many of the temples. Where the long lines of the lintols or architraves, pediments and stylobates (or bases) would appear to sag if built level, these were formed with slightly convex lines. Vertical features were inclined inwards to correct the tendency to appear to fall outwards at the top. The corner columns were increased in thickness to counteract the tendency to appear thinner when seen against the sky, this being lighter than the dark background of the solid cella wall. (See Fig. 11.)

The Greeks developed the 'orders of architecture', the Doric, Ionic and Corinthian. An 'order' consists of the column or post including a base and a capital and the entablature or lintol over. This entablature is divided into the architrave, surmounted by a frieze and capped by a cornice. The proportions of these separate parts vary in the different orders, as do the mouldings and decoration.

The Doric Order is the oldest, plainest and sturdiest. It is considered by many to be a direct derivation from a timber prototype. This theory has been disputed by many eminent historians, notably the Frenchman Viollet-le-Duc, but it seems probable that the theory is sound. The Doric column has no base but stands directly on a platform or stylobate, usually composed of three steps. The shaft is circular and divided, as a rule, into twenty shallow flutes separated by sharp arrises, and provided by an outward curvature or entasis to counteract the hollow appearance of straight sided columns. In early work this is excessive, but in the Parthenon (438 BC) the right degree of curvature was provided. The column is surmounted by a capital comprising an abacus (a flat square slab supporting the entablature or lintol), echinus (a convex or ovolo moulding below the abacus) and annulets (small stone rings around a

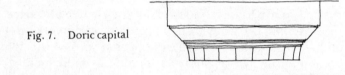

Fig. 7.   Doric capital

column). The profile of the echinus varies with the date of execution, earlier examples being fuller in profile than later, flatter examples. Annulets vary from three to five in number and are placed below the echinus to form a stop to the arrises of the flutes. The entablature has three main divisions: the *architrave* which corresponds to the lintol and is the main structural element spanning the opening, and the *frieze* which incorporates triglyphs (vertically fluted stones representing the ends of prototype wooden beams) separating metopes or square spaces sometimes filled with sculpture (see the Elgin Marbles in the British Museum). Crowning the whole entablature is the *cornice,* moulded and decorated, representing in stone the feet of sloping rafters and a roof constructed in timber. The Parthenon on the Acropolis in Athens and the Temple of Apollo on the island of Delos (300 BC) are very fine examples of the Doric order.

The Ionic order is remarkable for its scroll or volute capital. The columns are taller and more graceful than Doric, and usually incorporate twenty-four flutes separated by fillets and not by sharp arrises. The column is provided with an elegantly moulded base, but without a square plinth. The capital consists of a pair of spirals or volutes to the front and back connected by a 'cushion' and is provided with an echinus moulding carved with the 'egg and dart' decoration. The entablature, as in Doric, comprises three parts, the architrave usually in three parts also representing superimposed beams, the frieze often ornamented by a band of continuous sculpture and a cornice with continuous dentil ornament reminiscent of the ends of squared timbers. Two of the best known examples are the Propylaea (437 BC) and the Erechtheion (420 BC), both in Athens.

Fig. 8.   Ionic capital

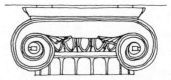

The Corinthian order is even more ornate than the Ionic. The columns are much taller, and the distinctive capital bell-shaped and adorned with two tiers of eight acanthus leaves surmounted by a curled leaf or calyx, from which spring volutes supporting the angles of the abacus. The abacus is usually moulded and concave in plan on each face. The entablature bears a general resemblance to Ionic with its triple division, the Corinthian mouldings having additional enrichment. Corinthian was little used by the Greeks, but the charming Monument of Lysicrates (335 BC) in Athens is a good example of this order.

The Romans utilised the columnar and trabeated (post and lintol) style of the Greeks, but joined to it the arch, the vault and the dome. Greek buildings were normally single storey but the Romans required structures of several storeys and the orders, attached and superimposed, were chiefly used as decorative features of little constructional significance. The Colosseum in Rome is a good example of the union between

Greek and Roman constructional systems, where the piers
between the arches on the different storeys are strengthened by
semi-attached columns acting as buttresses. Corinthian was the
order most favoured by the Romans, as in the Maison Carrée
at Nîmes in France (AD 200) and the Pantheon in Rome (AD
120), although Ionic is used in Hadrian's Villa at Tivoli.
Perhaps the order most employed by them was the Tuscan, a
simplified version of Doric with a plain, unfluted column and a
simple entablature. Columns of this order, devised by the
Romans, were economical to manufacture, and may be found
all over the Empire. Columns and bases of this order may be

Fig. 9.   Corinthian capital

seen at the Roman Palace at Fishbourne, near Chichester in
Sussex.

Early Christian architecture developed and incorporated
late Roman work in the provision of churches for use by the
adherents to the new faith. Most of the columns were salvaged
from Roman buildings, which had either fallen into ruin or
were purposely destroyed for their materials. The columns
were either incorporated with classic entablatures or, in later
work, with semicircular arches. The Basilican Church of S.
Maria Maggiore in Rome incorporates smooth Tuscan
columns with Ionic capitals and a richly decorated frieze.
Sometimes the orders were intermingled, the columns being
brought up to uniform height by the addition of new pieces, by
the provision of double bases or by the omission of base

mouldings. The use of column and entablature did not outlive the supply of Roman material and the almost total eclipse of the structural system was foreshadowed by the evolution and expansion of Byzantine architecture.

Christian architecture of the 4th to 6th cents AD marks the end of honesty in post and lintol construction. Revivals occurred, principally with the Renaissance and its later developments in the Georgian and Regency styles in England, and the New England style of North America. These were, however, styles and not structural systems. Although the posts and

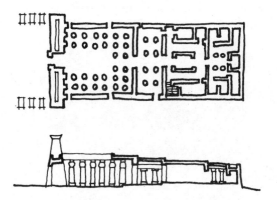

Fig. 10. Temple of Khons, Karnak. Plan and section

lintols were often loadbearing, the aesthetic beauty of Greek orders was applied as a veneer. Brick columns were faced with stucco and marbled, painted, or finished with scagliola. Only in the use of cast iron was there honesty. Many examples of cast iron columns can be found in Kent and Sussex, where the products of the Wealden ironworks were used with great effect in the simple Georgian buildings which enhance many of the villages and county towns of this part of England. Fine examples are Dixon's Ironmongery Emporium in the centre of Ashford, Kent, and a delightful row·of small shops in Hawkhurst. The Pantiles in Tunbridge Wells also exhibit fine examples of cast iron columns. The Romanesque round arch, developing into the pointed arch of high Gothic, dealt the

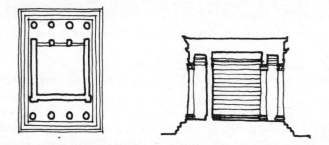

Fig. 11.   Temple of Athena Nike, Athens

death blow to the columnar and trabeated constructional techniques until the development of materials and structural theory in the late 19th cent. revived the earlier simple framed structures.

CHAPTER TWO

Corbel Domes and Vaults

Domes were constructed in the Middle East as early as 5000 BC. Simple structures of pisé de terre have been excavated at Nineveh which show distinct traces of the springing of vaulted roofs. Simple beehive structural forms have survived in this part of the world to this day and this form of construction was used for the burial chambers with domed stone roofs found in

Fig. 12.   Corbel vaulted tomb at Ur.
3000 BC

Crete and called 'tholoi'. Early examples were discovered in Ur, in the vaulted tombs which date from about 3000 BC. In Egypt, the corbelled roof of the Sanctuary in the Temple of Seti I (1350 BC) completed by Rameses II is another good example carried out in brick. Here, each horizontal course projects beyond the one immediately below, the soffite being worked by chisel into the smooth form of a vault. At this period there was very close inter-communication between Crete and Egypt, and it is not surprising to find architectural features common to both cultures. The Treasury of Atreus or, as it is sometimes called, The Tomb of Agamemnon, is 14·78 m (48½ ft) in diameter and follows the general design of the earlier 'tholoi'. The doorway is similar to the later work at Mycenae, the corbelled stones over the lintol forming a triangular opening which relieves it of its load. Internally, the dome is constructed of stone, each course corbelled out over the lower and chiselled to the smooth curve of the domical shape.

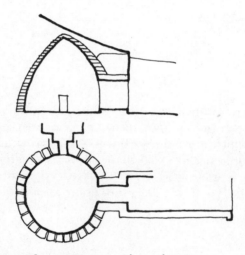

Fig. 13.   Treasury of Atreus, Mycenae. Plan and section

Another interesting example of a corbelled vault is to be found in the partly subterranean passageway which gives access to a well outside the walls of Mycenae (1400 BC). This city incorporates in its wall the Lion Gate, similar in construction to the entrance to the Treasury of Atreus, but with the triangular opening over the lintol, enclosed by the corbelled relieving arch, filled by sculptural representations of two lions flanking a tapering column. The great city of Tiryns, contemporary and rival of Mycenae, incorporates corbelled galleries in the construction of its massive cyclopean walls, and the

Fig. 14.   Gate of Lions, Mycenae

Tomb of Clytemnestra follows the same constructional pattern.

In countries where timber was not available for centering, an otherwise indispensable adjunct for arch construction in early cultures, a method was evolved to overcome this difficulty. Bricks forming the arch were laid so that each ring inclined upon the previous, held in position by the adhesion of the mortar and friction between the sloping planes. A brick vault from the Palace of Sargon at Khorsabad in Persia (800 BC) is an early example. It is probable that the arched vault of the Palace of Ctesiphon near Baghdad (AD 550) is the most daring example ever constructed. Spanning about 25 m (83 ft) and over 30 m (100 ft) in height, this structure is the largest unreinforced brick vault in the world.

The Byzantine builders of the 5th and 6th cents AD used the system of the inclined brick corbel for the construction of

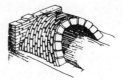

Fig. 15.   Brick vault at Khorsabad.
           800 BC

vaults and domes in preference to the Roman methods described hereafter. Although there had been domed structures before the Emperor Justinian, these domes were merely lids placed on cylinders whose thick walls had no difficulty in supporting their weight (the Pantheon). The problems of domical construction appeared when the dome was placed on a square base. This was not a difficult structural problem so long as the area to be roofed was small. A stone placed across each corner was sufficient to form an octagonal base from which the dome could be constructed. By replacing the stone

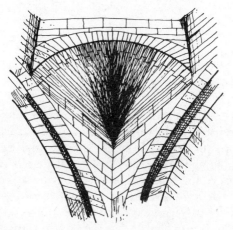

Fig. 16.   Squinch

with an arch, linked to the corner with a short length of vaulting, a constructional feature known as a 'squinch' was formed (S. Fosca, Torcello). The problem increased when, instead of a cube of stone, the dome was to be placed on four arches. The problem of thrust became paramount, those of the dome tending to force the supporting structure outwards.

The solution of the problem was the invention of the pendentive, which replaced the squinch at the corners of the square supporting the dome. A pendentive is a spherical triangle, a section of a dome which, starting at the corners of the square from the capitals of the supporting columns, rises between the perpendicular arches until the four meet to form a circle. This circle rests on both the keystones of the arches and also successive courses of the pendentives, and provides a base from which the dome can be constructed. The thrust of the dome is thus distributed over the whole surface of the pendentives and arches to the four supporting columns. It was necessary to reinforce these to prevent collapse and this was effected in several ways. First, by regarding each column as the angle of two walls and constructing four naves in the form of a cross with vaulted roofs, or half-domes if the naves were apsidal in form. Secondly, the arches of the principal dome could be buttressed by four smaller domes, in turn reinforced on the exterior by buttresses. Whichever structural system was

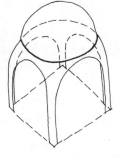

Fig. 17.   Pendentive

adopted a cruciform plan was automatically produced, often provided with halls at each of the four corners whose vaults assisted in stabilising the structure, and with the addition of an apse at the East end and a narthex or vestibule at the West.

While columns and piers were generally of stone, the basic material of Byzantine architecture was brick, set in very thick beds of mortar which enabled corbelling to be more readily carried out. This also ensured the minimum of cutting to provide the smooth face necessary to receive the mosaic with which these structures were internally finished. Two churches of this period illustrate the constructional techniques used and each is a masterpiece in its own right. S. Vitale at Ravenna in Italy (AD 526–547) differs from the usual Byzantine plan in that the dome is supported by an octagon, each facet with lofty

Fig. 18. S. Vitale, Ravenna

arches carried by columns provided with lateral support by eight half-domes, acting as buttresses for the central structure. These half-domes are in turn buttressed by further vaults in two tiers joining them to the strong outer walls.

The problem facing the architects of the Hagia Sophia in Constantinople (AD 532–537) were much greater. They had to raise a dome 23 m (75 ft) in diameter a similar distance above the ground and place it on four arches, to be the centre of a further series of vaults. The plan of the church was a rectangular basilica, its length twice its width. The main dome was stabilised by half domes provided between the side arches, of

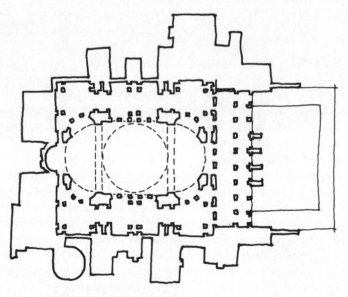

Fig. 19. S. Sophia, Constantinople. Plan

the same diameter as the main dome, and buttressed by three others, with the Eastern replaced by the apse and the West by an internal porch. The main axis of the building was 76 m (250 ft) long. Lateral stability was provided by enormous interior buttresses placed against the walls behind the pillars. Unlike Western cathedrals, the church was not interrupted by bays or side aisles and the width is maintained throughout the length of the building.

The methods employed by Byzantine architects spread mainly through the Greek and Slav world. In Western Europe, the province of Aquitaine in France possesses a magnificent

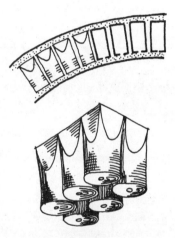

Fig. 20.   Bottle bricks or 'cones' built into corbel domes to reduce the weight

group of churches vaulted with a series of stone domes, supported on pendentives. These domes roof the choir, nave and transepts as well as the crossing and there can be no doubt that these are of Byzantine origin, scattered as they are along the Roman road from Rodez to Cahors and then Saintes through Perigueux and Angoulême. Perhaps the most famous of these churches is St Front at Perigueux.

The cathedral of Cahors was rebuilt by Bishop Geraud III (1109–1212), after his return from pilgrimage to the Holy Land. He must have seen many Byzantine churches during his travels and his model for Cahors might well have been the

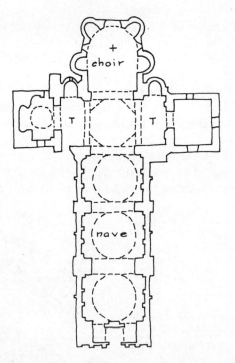

Fig. 21. Angoulême Cathedral. Plan

Church of the Holy Apostles at Constantinople, now destroyed. At Cahors, two great domes on pendentives balanced only at their corners provide an original conception by abandoning the constructional methods of Romanesque building current at the time of its construction. Nowhere else in Western Europe are structures of this type to be found dating from this period, and with the completion of the building of these churches the technique of the corbelled vault and dome died for all time.

CHAPTER THREE

Round Arch and Vault

The simple method of spanning openings with a flat stone lintol was employed in Egyptian building and later copied by Minoan, Hellenic and Hellenistic builders. The problem of spanning openings of any size was considerable, the slab over the Lion Gate at Mycenae, weighing between twenty-five and thirty tons, being used to bridge a span of little more than 3 m

Fig. 22.   Temple of Apollo, Delos

(10 ft). By resting inclined slabs against each other larger spans could be bridged, but the abutments had to be of sufficient weight to resist the lateral thrust. Where these were in solid rock, such as the entrance to the temple of Apollo at Delos, there was little problem. Here ten slabs, five on each side, span a distance of about 6 m (20 ft).

A small semi-circular brick arch built about 1400 BC at Ur in Mesopotamia is said to be the earliest example of the arch as an architectural element. The arch spans 0·8 m (32 in) and is built of kiln-burnt voussoir bricks forming an opening through a wall 1·6 m (5 ft 4 in) thick. Another example dating from about 1300 BC is at Susa in Persia.

The Greeks, like the Egyptians, preferred the use of the flat lintol, but the Etruscans of Northern Italy built arches with close fitting worked stone voussoirs, constructed with dry joints. This technique was adopted by the Romans who used the semicircular arch for spanning openings in walls, for relieving arches and for vaulting. Special, tapered, arch tiles were produced which were in effect thin voussoirs and examples of these were found at the excavations at Silchester. A

Fig. 23.   Brick arch at Susa. 1300 BC

technique developed by the Romans was that of the joggled voussoir, which assisted erection by preventing stones from sliding on each other and locating them in position. Examples of this constructional technique can be found in the bridge over the River Pedroches near Cordova in Spain and in Diocletian's Palace at Spalato in Yugoslavia (303 BC–AD 5). Occasionally a flat arch was provided, but this always had a relieving arch to reduce the load. The round vault which is such a feature of Roman buildings may originally have evolved as an imitation of a natural grotto. These grottoes played an important part in the Greco-Roman mythology, and were held to be the dwellings of supernatural beings; the gardens of Roman villas often included artificial grottoes of religious significance.

Roman basilicas were rarely provided with vaulted roofs or apses before the Augustan period. Certain buildings designed for essentially practical purposes, such as the Thermae or public baths, were roofed with barrel vaults completed by the semi-dome of the apse (Stabian Baths at Pompeii). The vaulted halls of the Augustan period are of modest dimensions and the earliest example of a groined vault also dates from this period. When the Domus Aurea was built to Nero's orders, in the middle of the East wing was erected an octagonal hall roofed by a dome. This was not supported on a continuous wall but on eight pillars set in an open square. Unfortunately this building has now completely disappeared, but it can be compared with the Pantheon rebuilt by Hadrian about AD 120.

The Pantheon is essentially a vast brick rotunda, 43 m (142 ft) in diameter and crowned with a dome reaching to the same height. The wall of the supporting drum is relieved by eight

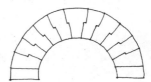

Fig. 24. Roman concrete with brick relieving arch

Fig. 25. Stone arch with joggled voussoirs

niches; in the front of six of these are two columns which appear to support the dome. In fact, the thrust is diverted onto the massive pillars by relieving arches.

The earliest example of the use of a groined vault over a large area may be found in the Baths of Trajan at Acholla. The combination of a groined vault and two barrel vaults was replaced by the juxtaposition of three groined vaults in the Baths of Trajan at Rome and this is again found in the Basilica of Maxentius. In this case it is balanced by two aisles, each counterbalanced by three barrel vaults at right angles to the main axis. The barrel vaulted hall, closed at one end with an apse and with its longer sides relieved by niches, was used by the Romans in the 2nd and 3rd cents AD for the shrines of several of their larger temples (Temple of Venus in Rome).

The construction of most of the vaults was carried out in concrete, often relieved by brick arches integrated into the mass, which provided support for the shuttering and construction joints to relieve shrinkage. The vaults over some of the smaller bath-houses were often constructed with arches formed from hollow voussoirs, or with solid masonry supporting two thicknesses of tile, separated to form a void. The voids and hollow voussoirs were heated by the furnaces and this, raising the surface temperature on' the underside, helped to prevent condensation. While, as we have seen, stone was employed for arches, and probably the finest examples are in the Pont du Gard near Nîmes in France, brick was the most common material used for this purpose, especially in later work. This provided a finished face for the backing work in concrete, and magnificent examples of round brick arches are to be found in the Baths of Trajan in Rome.

Fig. 26.   Flat arch with relieving arch over

The semicircular arch was an important architectural feature of Early Christian and Byzantine architecture. Many of the early buildings utilised Roman elements and re-erected columns and masonry obtained from derelict or ruined structures of the old Empire: semicircular arches for arcade orders, windows and doors (St Appolinaire at Ravenna), apses with semi-circular domes over (St Sabine in Rome). Domes and barrel vaults were constructed in stone and brick, sometimes combined. While most of the Early Christian basilican churches were provided with wood roofs of pitched construction, the octagonal baptistries constructed in great numbers, and many of the Byzantine churches, were either vaulted or domed. The domes and vaults are semicircular, the domes usually supported on pendentives when constructed over a square plan.

Romanesque architecture, covering the period from the fall

of the Roman Empire to the beginning of Gothic, depended primarily on the round arch and vault in its structural techniques and reached its zenith in the year 1095. Romanesque architects had little mathematical knowledge, they were practical men working with an acquired skill. To achieve the cruciform plan used in their churches simple geometrical shapes were employed – squares, rectangles, circles and semicircles. Often their plans were remodelled and revised as the work progressed. On occasions they even demolished work if they felt that the alterations envisaged would benefit the overall composition.

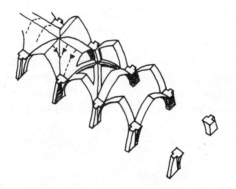

Fig. 27. Construction of a cross vault without ribs

Ottonian architecture opened the way to Romanesque art. This had developed under the successors to the Carolingians in Aachen in Germany, and provided patterns for monumental buildings with spacious interiors, symmetrically placed towers, regular crossings, interior galleries and alternation of arcade supports. The principle of the continuous transept forming a barrier to the nave and separating it from the apse was a favourite device of the German architect and the regular crossing first appeared in the church of St Michael at Hildesheim. Here both nave and transept are the same height, meeting at a narrow crossing reduced to a square by four fine equal arches.

The great abbey of Cluny in France is the cornerstone of Romanesque design. The church was begun about AD 950, basilican in plan, with an unvaulted nave of seven bays flanked by aisles and a narrow transept. The choir terminated in a central apse flanked by two smaller apses projecting from a straight East end wall. The nave was preceded by a narthex or porch. The church, 45.5 m (150 ft) long, was too small by the 11th cent. and a new church was constructed to the North, with a nave of eleven bays, balanced by double aisles and transepts, and an ambulatory with small radiating apses. A similar plan, in a simplified form, was provided for the Clunic Priory Church of Lewes in Sussex.

Very rarely is detailed information to be found on the careers of laymen in Romanesque writings. Usually the name of the patron was considered sufficient; the slightest trace of personal gratification was barred, and total involvement in the building was all-important. Exceptions to this rule are therefore interesting. For the reconstruction of the Romanesque Cathedral of Compostella in Spain, it was recorded that the work was entrusted by Bishop Diego Pelaez to qualified representatives of the cathedral chapter. The records show that the general direction of the work was carried out by a monk – Bernard – treasurer of the chapter, assisted by a deputy named Robert and a staff of about fifty stonecutters. Two centuries later information of this kind would be recorded in the accounts and contracts prepared for mediaeval building work.

The original Norman/Romanesque in England was aesthetically severe. This character may still be admired unaltered in the naves of several churches: Binham Priory in Norfolk, and Romsey, Selby and Waltham Abbeys. The walls of these last three churches are powerfully articulated with arcaded galleries and wall passages. The sturdy rhythm of the great pillars with their vertical emphasis must have foreshadowed the future rib vault. It was not, however, until towards the end of the 11th cent. that its forerunner, the groined vault, was more widely used throughout the Romanesque world, despite its complex construction.

There is a basic difference between English and Continental Romanesque architecture. The English style did not arise from a spontaneous act of creation or the inspiration of instinctive genius. It was a foreign art imposed from abroad. Some of the finest Norman churches were built in England. They share the robust proportions of Ottonian Architecture of Germany with superb tripartite elevations, great round arches with rich round mouldings, triforia or wide galleries and an upper gallery rising above the pillars in front of clerestorey windows.

In 1093, William Bishop of Durham began a new cathedral to replace the original Saxon church. The plan comprised a nave of eight bays including two Western towers, with transepts of three bays and a choir and aisles of similar dimensions.

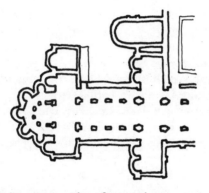

Fig. 28.   Lewes Priory, Sussex. Plan of East end

These latter terminated in apses. As the walls of the North transept neared completion the bold decision was taken to cover it with a vault. The wood roof of the South transept was removed, and both this and the nave were vaulted in stone. The choir had originally been rib-vaulted. This matched the side aisles which today retain their original work, the choir having been rebuilt in the 13th cent. (See Fig. 37, p. 45.)

In the choir aisles the vault compartments are divided by transverse arches which spring from both the columns and the piers of the alternating arcades. By stilting the transverse ribs it was possible to retain the familiar round arch. In the high vault of the nave, the transverse arches run only from the piers, the diagonal ribs being supported by corbels above the columns, each compartment between the piers containing two sets of diagonal ribs. In these oblong spaces it was impossible to use round diagonals springing from the same level as the transverse arches. The problem was solved by the use of the pointed arch for the transverse ribs and the wall arches enclosing the clerestorey. It was also found that these pointed arches could be raised to any height, thereby presenting a level crown to the vault while still setting the height by the semicircle of the diagonal.

Much has been written to prove that the ribs of a vault were not structural and that a groined vault was as strong without them. At Durham the relationship of the ribs to both the supports and the buttresses is clearly articulated and must be the result of calculated planning, apart from the advantages of the stone ribs in the erection of the vault. The new method evolved at Durham was soon copied. After the collapse of the central tower at Winchester in 1107, the bays of the transept aisles were revaulted with ribs with a semicircular curve and the half roll mouldings of Durham. The choir aisle vaulting of Peterborough is similarly treated and at Lindisfarne and Selby the work was probably carried out by the Durham masons' yard.

The yard at Durham was employed on several important works. Our knowledge of the organisation of such yards is incomplete; masons were itinerant rather than local, and the main quarries were probably the principal training grounds. Some workmen may have come to Durham with the loads of Caen stone. If the Durham style is indicative of their employment, these masons must have travelled South as far as Essex, and others as far as Orkney. Lindisfarne Priory was founded as a cell of Durham and a Durham monk was in charge of the administrative side of the work. The abbey church of Selby was begun by Abbot Hugh de Lacey (1097–1123) who is described as 'architectus'. Here the plan is similar to Durham and most certainly Durham masons were employed. From the first, the quadripartite vault was envisaged. The Abbey of the Holy Cross at Waltham had special ties with Durham, many of its estates having passed to the Northern bishopric, and the properties of the Durham elevation are retained with clere-storey passage, alternate columns and compound piers.

The next great developments, culminating in the evolution of the pointed arch and what we know as Gothic, took place in Normandy. Sometime after 1106 rib vaults appeared in the Chapter House at Jumièges, and in the Abbey of St Etienne at Caen the sexpartite division of the vault was attempted. This system was not brought to England until the last quarter of the 12th cent. for the rebuilding of the choir at Canterbury.

Early Romanesque architecture in England is usually known as Anglo-Saxon. In this, round arches were used to span openings, in conjunction with other methods, often relieved by

Fig. 29.   English Romanesque doors and windows

superimposed arch mouldings incorporating distinctive im-posts at the springing. The jambs were usually square and where arched openings were provided in plurality, inter-mediate support was given by the use of turned stone balusters (Earl's Barton, 980). In later work, after the Conquest, semi-circular openings were frequently formed with square recesses or orders to the jambs. In early work the windows were usually small, narrow and deeply splayed internally. In later examples both doorways and windows increased considerably in size and were often richly ornamented with zig-zag and beak head ornaments (Iffley Church, Oxford). Sometimes the panel en-closed by the semicircular head, called a tympanum, was richly carved with sculptural subjects (Barfreston, Kent). Late

Romanesque arcades were similar to the windows and doorways. In addition to the ornamentation mentioned, billet,
nailhead and roll mouldings were used. Vaulting ribs were
usually decorated with roll mouldings with bold undercutting
(Durham Cathedral).

As early as the 11th cent., the Italians were looking back to
the glories of Augustan Rome. Tuscan Romanesque moved
uneasily in the Gothic style, never properly understood or
liked by the Italians, and by the 14th cent. the writings of
Petrarch opened their ears to a new philosophy of humanism –
the human as opposed to the abstract theology. In the 15th
cent. the classical writers of antiquity were rediscovered and
the Renaissance emerged in Italy.

The architect for the rebirth of the antique was Brunelleschi,
a Florentine born in 1377. The freshness and originality of his
San Lorenzo (1425) contrasts with the Gothic ribbing supporting the cupola of Florence Cathedral built by Brunelleschi
between 1419 and 1436. Perhaps the finest building by this, the
first true architect of the Renaissance, was the Foundling
Hospital, also in Florence, with its classical loggia incorporating Corinthian columns, supporting semicircular arcades, its
wide entablature and rectangular windows with triangular
pediments. Round-headed windows, barrel and groined
vaults, semicircular domes were all borrowed and utilised from
Roman sources. The style spread across Europe and, coincid-
ing with the Reformation in England, was manifest in a crude
form in Elizabethan and Jacobean buildings in the 16th and
17th cents. The great English protagonist was Christopher
Wren, who used semicircular arches, barrel vaults and domes
in his work at St Paul's. The classical movement continued
throughout the 18th cent., producing much fine but often
theatrical domestic work, best known in the charming semicircular headed doorways and Venetian windows which characterised the work of the Adam brothers.

Although most arched and domed roofs had, in the past,
been constructed of brick and stone, new materials were often
used to simulate the effect. At St Paul's, Wren used timber
supported on a brick cone to provide the outline of the dome.
Internally, the finish was painted plaster. Earlier in the 16th

cent. a French architect, Philip Delorme, proposed to construct roofs and domes with a series of arched timber ribs in place of trusses. The ribs were of planks in short lengths, placed edgeways and bolted together breaking joint. The method of construction was used in the erection of a dome of 36·5 m (120 ft) diameter for the Halle au Blé in Paris. Another smaller example was that roofing the Pantheon Bazaar in Oxford Street, London.

The system was, however, wasteful of material and labour. In 1817 a French military engineer, Col. Emy, proposed an improvement on the system and in 1825 he erected a roof of 20

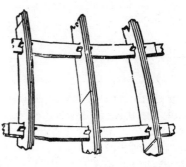

Fig. 30.   Curved ribs roof frame

m (65 ft) span at Marac near Bayonne. In this structure the ribs were formed of planks bent round on templates to the required radius, held in position by iron straps and radiating struts. This exerted no thrust on the supporting wall, the whole weight being thrown onto the wall at the feet of the ribs (Fig. 31).

One of the largest arched roofs ever constructed was that provided over the Imperial Riding School in Moscow in 1790. The principal feature of this roof, spanning 71·5 m (235 ft), was an arched timber beam formed of three pieces of timber notched to prevent them sliding over one another. The ends were kept from spreading by means of a tie beam, the two connected by suspension struts and diagonal braces.

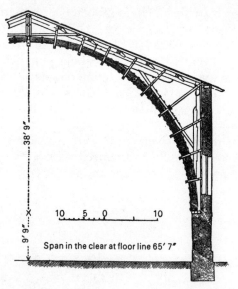

10　　5　　0　　　　　10

Span in the clear at floor line 65' 7"

38' 9"

9' 9"

Fig. 31.　65 ft span ribs

In 19th cent. roofs of round arch construction were of cast iron or steel, and were erected over many railway stations. These structures were often fine in conception and execution. Laminated timber trusses of similar profile have today become common for cheap roofs where large areas, unencumbered by supports, are required. Where the structure has to withstand corrosive conditions, shell concrete barrel vaults are often provided, thus bringing the semicircular arch back full circle to one of its principal uses, to provide a roof over the humid atmosphere of the Roman bath.

CHAPTER FOUR

Pointed Arch and Vault

Whether the pointed arch originated in Syria, where examples date from the 6th cent. AD, or was a natural development from Romanesque, is open to question. Its development became general in Europe, except in Italy, by the middle of the 13th cent. and is considered to be due in the main to sound structural reasons evolved, as we have seen, in the new vaulting at Durham Cathedral.

It is also possibly correct to say that the Gothic system of construction came into being with the building of Abbot Suger's Abbey of St Denis near Paris. The building only survives in rudimentary form, its original foundations having been unequal to their task, but the rapidity with which its ground plan was adopted shows how inspired a solution this was. (See Fig. 27, p. 31.)

The constructional principle of Gothic depends on relieving the external walls almost completely from vertical and horizontal stresses imposed by the arch. Now, the ribs, previously bonded into the keystones, were constructed to provide independent statically balanced crossties, independent also of the infilling of the vault itself. The keystones were carried by a series of cross ribs to which they were fitted. The vertical and horizontal stresses were taken almost entirely by the cross ribs

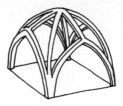

Fig. 32. Construction of ribbed vault
(See also Fig. 34)

and diverted to the four corners of the bay where the vault met
the walls. To provide further directional control here, upright
shafts or clusters of shafts diverted the vertical forces to the
floor by way of the arcade piers. Lateral thrust, on the other
hand, was conveyed to the exterior of the building where
buttressing was provided to support the weight of the vault.
Around the outer perimeter of the building heavy buttresses
extended upwards into pinnacles, absorbing the horizontal
stresses from the vault by collecting these through flying
buttresses. Internally, the new system, providing as it does for
the transmission of vertical loads direct to the foundations by
way of clusters of shafts and piers, relieves the walls altogether
of all load. These can then be ignored from a constructional
point of view, being replaced by openings. At St Denis the idea
of choir chapels, already to be found in Romanesque churches,
was developed. The chapels stand so close together that they
merge – even the side walls are open, giving the appearance of
a double ambulatory. Only the curved outer wall and window
remain today. Without the Gothic or pointed arch, this solution to the problem of enclosing space was impossible. Only
the versatile and elastic system of cross ribs and pointed arches
made it possible to preserve the same ceiling height in the

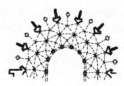

Fig. 33. Abbey of St Denis, Paris. Plan of
radiating choir chapels

ambulatory and chapels and yet cover so many spaces of varying plan. The monastery church of St Denis must be regarded as one of the most important in the development of cathedral architecture. Suger became abbot in 1122. He was Regent of France while Louis VII took part in the second crusade. The church was the burial place of the kings of France and was, in fact, completed in a very short space of time. The original church constructed in Carolingian days had had two basic faults: the entrance was too narrow for the crowds of pilgrims, and there was insufficient space around the altar for the relics to be properly displayed. The new West front with its three bays incorporating large and imposing doorways was completed in 1140. Abbot Suger then turned his attention to the choir. Retaining the old crypt, to place the new choir on a higher level so that the relics might be more easily seen, a double ambulatory with seven chapels around the main apse was constructed, facilitating the circulation of pilgrims. Between the external supports of the radiating chapels, the voids were filled in with stained glass.

The organisation behind the building of a cathedral is almost as remarkable as its actual construction. The construction usually covers several generations of builders, during which the original idea must be retained while at the same time advantage must be taken of new techniques and motifs which develop. The organisation was directed by a master builder who, having completed his apprenticeship as a mason, would have travelled far and wide as a journeyman, probably all over Europe, working on various buildings and assisting master builders in their work. His sketch book would reveal his experience, new constructional forms and ideas. The relationship of the master builder to his successors was controlled by strict craft mysteries. The best craft mason would be selected as assistant, and he would be instructed not only in the calculations and the preparation of fnaterials, but in how to handle men and, especially, the highly placed patrons of the project. He also learnt how to interpret the theological requirements and translate these into architectural terms and concepts. This craft experience was handed down from generation to generation and found practical expression in the expertise of the

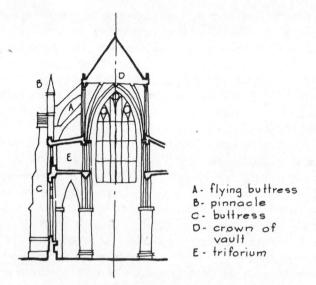

A - flying buttress
B - pinnacle
C - buttress
D - crown of vault
E - triforium

Fig. 34. Westminster Abbey. Part section showing system of flying buttresses to withstand the thrust of the stone vault

master builders. These men were among the great personalities of their time, friends of kings and bishops, working closely with them in the financing and planning of buildings.

The primary form of the church as we know it today was developed under Christianity in the late Roman period. The basic ideal of cathedral architecture is the basilica, developed in the time of Constantine from the Roman secular building into a Christian church. The plan was essentially longitudinal, with three or more parallel aisles, separated by rows of

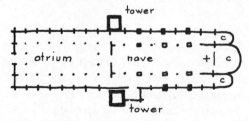

Fig. 35. S. Ambrogio, Milan, 1088–1128. Basilican church with atrium

columns, the wall of the central aisle rising above the others to permit the introduction of windows. The axis was usually West to East from the entrance to the Eastern apse, and sometimes one or two transepts crossed the nave and provided additional space.

Every era of mediaeval building interpreted this basic plan in different ways. Romanesque is characterised by the addition of closed units one to another – rectangular bays each with its own separate groined vault – the transeptal crossing adding separate stops to nave and chancel, the crypt with its usual effect of raising the floor level of the chancel, and separate galleries or triforia over the side aisles, often with their own altars. Gothic is different, the congregation being gathered into one united body; the transepts no longer project so far and, their height being equal to that of the nave and choir, the crossing is no longer an independent space. In this, Gothic returns to the single directional entity which marked the early Christian basilican concept. (See Fig. 38, p. 46.)

While the French rapidly advanced to the realisation of spatial unity, English cathedrals continued to adhere closely to the Norman tradition of separate plan units. A typical English cathedral is Wells in Somerset. The West elevation is broader than the building itself, and while each aisle has its separate entrance, the supporting towers stand to one side, producing an elevation of monumental quality. The cathedral rises from a wide expanse of lawn, utterly foreign to the French cathedral, always set down in a huddle of houses. The entire elevation of Wells above ground is built up on the translucent wall principle, the wall itself hidden by a confusion of arcades, with tall narrow arcades of the upper wall between the buttresses giving an impression of soaring grace. The greater part of the West elevation was erected between 1220 and 1239, the upper parts of the towers between 1367 and 1424.

Internally, the nave stresses, in contrast to French work, the independent horizontal line. Here the three divisions of the elevation, arcade, triforium and clerestorey, are in direct contrast to the vertical relationship between the parts and nowhere do the numerous responds of the ground level piers contribute to the support of the vault. This is all in direct contrast to the

French ideal where the responds of the vault reach down through the composite piers to the floor. The plan makes it clear that the choir, following the crossing and the transept, is an independent church, and even further East there are the lateral additions of two chapels forming a second and smaller transept. This arrangement of plan also occurs at Canterbury, Salisbury and Lincoln. At the East end of the cathedral is the Sanctuary with its projecting Lady Chapel, the whole providing a manifold spatial dimension designed to serve the ritual of English liturgy.

Outside the grand plan itself, the English cathedral usually includes an extensive cloister with an inner courtyard, bordered at Wells by the Bishop's Palace. To the North,

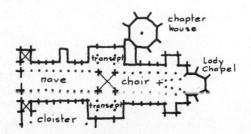

Fig. 36. Wells Cathedral, Somerset. Plan

alongside the choir, stands an octagonal two storey chapter house with a nearby gateway leading to a row of ecclesiastical dwellings, 'Vicar's Close'. All this converts a complex structure into an independent settlement.

One of the principal features of the Romanesque church was the division of the internal bay into three parts, and this, as we have seen, was perpetuated in many of the later Gothic structures. The triforium has always been an interesting element in church design, serving as it did initially the gallery void over the vault of the side aisle. As Gothic architecture progressed the triforium shrank until it disappeared, leaving the primary storeys of arcade and clerestorey. The clerestorey was of the utmost importance, providing light to the main part of the church, and was therefore usually a fine architectural feature.

Clerestorey windows were provided with passages from which the glass could be cleaned, and the treatment of these walkways provided further opportunity for fine architectural treatment. In the 11th cent. this usually took the form of a triple lancet opening which, as the width of the bay increased, was later discarded to allow the whole width of the bay to be filled with glass; the cill then became the cleaners' access (York Minster).

The main arcades of the 11th and 12th cents were low and supported on sturdy columns, usually round (Durham Cathedral). In the 13th cent. the height of the arch, due to the introduction of the pointed form, was raised and, in conjunc-

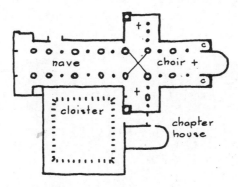

Fig. 37.   Durham Cathedral, *c.* 1180. Plan

tion with the introduction of the moulded column, introduced Gothic grace (Salisbury Cathedral). During the 14th and 15th cents the arcade increased in size and scale, bays became wider and roofs rose higher. The external walls were reduced in weight and volume, until they consisted only of the wall arcades which decorated the aisles and transepts of so many mediaeval churches (Lincoln Cathedral), and the buttress. The broad sturdy pilaster buttress of Romanesque builders, provided more for punctuation than for structural necessity, vanished with the invention of the Gothic buttress, built to withstand the thrust of quadripartite vaulting. Here the pressure was arranged to be concentrated on the centre lines of the bay structure where heavy masses of masonry were required. At first these buttresses were set back towards the top to obstruct the windows as little as possible, while the introduction of heavy pinnacles produced the necessary mass and provided the essential verticality which is the essence of Gothic architecture.

All this verticality needs some horizontal feature to tie the verticals together. In Gothic architecture this was provided by the plinth, gradually becoming more and more elaborate and providing a strong aesthetic foundation on which the building could stand. String courses were also provided, tying in the cills of windows and the pierced parapets to the general design.

The ultimate in Gothic pointed arch construction can be seen in the Late Perpendicular structures of King's College Chapel at Cambridge and Henry VII Chapel at Westminster

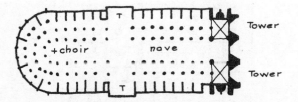

Fig. 38.   Cathedral of Nôtre Dame, Paris

Abbey. Henry VII Chapel (1503–1519) is a spatially autonomous structure built on such a scale as to appear almost larger than the Abbey itself. This is one of the most important examples of English Late Gothic construction, with its unity of space in place of the sub-divisions of previous structures. The whole conception is emphasised by the decorative effect of mullioned window and pannelled wall which rises up and through the fan vaulting as if petrified in movement. King's College Chapel, on the other hand, is a separate building conceived as a whole and brilliant in its execution. Begun in 1446 as an aisleless nave and choir of twelve bays, the present effect is spoilt by the Renaissance organ case dividing the chapel into two parts. The surfaces of the walls are richly pannelled and merge unobtrusively into the great windows of the clerestorey. The whole interior is enriched by a fan vault completed in 1512, which rising from the arcade ribs, floats serenely over the chapel. These two buildings are the climax of Gothic architecture in England. Within the space of time covered by two generations, after the dissolution of the monasteries by Henry VIII, the craftsmanship which had produced the exquisite work at Cambridge and the Abbey had died for ever. Hawksmoor, at All Souls College, Oxford, and in the Western towers of Westminster Abbey, tried in the 18th cent. to recapture the spirit and brilliance of the past, but in vain. Charles Barry and Augustus Pugin in their design for the new Houses of Parliament in the 19th cent., and many other architects of this period with their pseudo-Gothic churches and town halls of brick, terracotta and cast iron, illustrate the futility of revivalism without traditional craftsmanship.

CHAPTER FIVE

Structural Frame

Wood has always been man's most useful building material, and before the advent of advanced metallurgy, which enabled metals to be substituted, it remained the chief material for complete structural frameworks.

The type and quality of available timber depends on both the prevailing climatic conditions and the composition of the soil. In hot dry areas such as the Middle East, good timber has always been scarce, indigenous trees such as acacia, palm, tamarisk and willow providing only light timbers for building. The Greeks and Romans used, in addition to those mentioned above, oak, box, alder and fir. Changing climatic conditions led to the appearance of different varieties of timber and to successive changes in building techniques. Early structures may well have comprised a shallow pit with the excavated material arranged around the perimeter in the form of a bank, with central posts supporting a ridgetree which in turn supported the ends of rafters carrying the roof of turf or brushwood. Excavations at the Iron Age settlement at Meare near Glastonbury in Somerset, where a village of about 70 dwellings clustered in an area of about 1·5 ha ($3\frac{1}{2}$ acres) on an artificial island, showed that the houses of this period were 5·5–10·5 m (18–35 ft) in diameter with walls formed from posts driven into the earth, the spaces being filled with wattle, daubed with clay. In areas where timber was more readily available, walls of a more solid construction utilising split tree trunks or planks were built. This form continued in use up to Norman times in England and an example may be seen in the small church of Greenstead in Essex, where the Saxon nave is formed from split logs set on end on a modern stone plinth. The stave churches of Norway and Denmark are mediaeval counterparts of these early buildings, and farmhouses constructed of roughly squared logs were common in Scandinavia and Eastern European countries. Solid timber walls were very extrav-

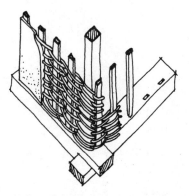

Fig. 39.   Iron Age wattle and daub construction

agant in material and where supplies of suitable timber were dwindling it was found to be more economical to construct a timber framework and fill this in with another material.

As we have seen before, the Romans were quick to seize on the practical advantages of any idea. In Romano–British domestic construction the framework of the Iron Age house was improved and adapted, the wall plate either being bedded on a dwarf wall or buried in a trench below floor level and cased in concrete. Variations of this technique occur in much mediaeval work.

It is likely that more buildings have been wholly or partly built of wood in the last 900 years than of any other material.

Forests were once abundant, and in the 13th cent. over sixty
genuine forests were described and named. Practically all
English trees were hardwoods and the timber universally used
for building was oak. Despite its tendency to warp and crack
and its susceptibility to beetle attack, it is very hard, close-
grained and durable. Wealden oak was famous and the Welsh
border counties comprised one vast oak forest. English oak is
not always suitable for joinery and from the 13th cent. oak and
some pine was imported from Germany for panelling and
joinery. After the middle of the 16th cent. oak in some areas

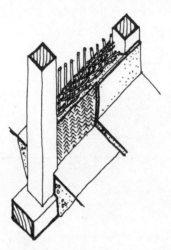

Fig. 40.   Romano-British wattle and daub construction

became scarce and consequently Baltic pine and fir was imported in ever increasing quantities supplemented, in the 19th
cent., by vast shipments of similar timbers from North
America.

Oak was not always seasoned before use. Air seasoning was a
long business taking several years, and water seasoning (im-
mersing barked logs in running water for a minimum period
of a year) was often used. This lack of seasoning accounts for
much of the warping and splitting found in oak timbers in
mediaeval construction.

In secular work, apart from military structures, wood was

preferred to stone not only on account of its lower cost but also for its ease of handling by relatively unskilled labour. The majority of English towns including London were, in 1600, entirely of timber framed houses. These were usually known as half-timbered, that is to say the frame was of oak, with the infilling of wattle and daub. Later, many of these framed structures were faced with tilehanging or clapboard. In some cases the perishable wattle and daub was replaced with brick nogging, often in a decorative herring-bone pattern. Framed

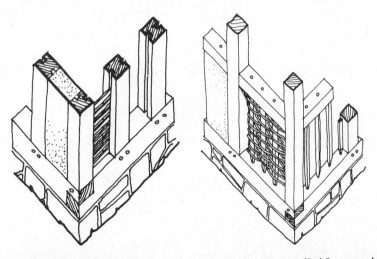

Fig. 41.   Close-timbered framework   Fig. 42.   Square-panelled framework

houses of this pattern occur in most areas where local stone is not readily available, although few houses have survived from earlier than the 16th cent. Today timber framed houses are mostly found in rural areas and small towns South-East of a line from Huntingdon to Yarmouth and West of a line from the Severn Estuary to Leicester, continuing West of the Pennines.

In the simplest timber structures there was no distinction between wall and roof. The framework consisted of pairs of tree trunks, their thick ends charred for preservation, set in the ground with their tops crossed and tied together about a ridge pole. In England the timber framed structure evolved from what is known as 'cruck' construction. Crucks were cut from trees with, if possible, a natural curve; sometimes a single tree would be split along its length to provide a degree of symmetry. The crucks were set up on a stone plinth or similar foundation at intervals, linked horizontally at the top by the ridge pole. Lower down, the crucks were tied by purlins with the interstices filled in with brushwood and thatched with straw or heather. These simple, single storey houses developed with the introduction of tie beams, which allowed the insertion of an intermediate floor. By lengthening the tie beams an improvement in headroom resulted from flattening the pitch of the roof and introducing vertical walls on the ground floor.

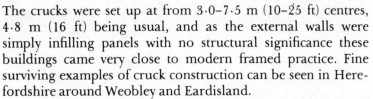

Fig. 43.   Cruck construction. Detail
through plate and purlin

The crucks were set up at from 3·0–7·5 m (10–25 ft) centres, 4·8 m (16 ft) being usual, and as the external walls were simply infilling panels with no structural significance these buildings came very close to modern framed practice. Fine surviving examples of cruck construction can be seen in Herefordshire around Weobley and Eardisland.

The length of a cruck framed house could be extended indefinitely, but in practice it was restricted in width and height by the size of timber available for the crucks. The problem was solved by the evolution of box frame construction which, although possessing local variations, had certain essential features. The cill was usually placed on a low plinth or raised foundation of stone, and strong upright posts or 'studs' were mortised into it and secured above in a similar way to a heavy beam or head piece. This timber either carried the floor joists or the rafters of the roof. The upper storey frequently projected on one or more sides and the overhang was known as a 'jetty'. This was probably provided to assist in protecting the lower part of the structure from storm water from the roof, no gutters being provided. To construct the overhang, the corner post of the upper storey was supported on a 'dragon' beam projecting diagonally from a support in the centre of the house, into which were tenoned the shortened floor joists on either side. The corner support under the jetty was usually formed by an outward curving piece of timber acting as a corbel and known as a 'dragon post'. This timber was usually moulded or carved. Braces were inserted to the corners to provide additional support and stability. In Lancashire and Cheshire a cove was sometimes provided under the overhang (Little Morton Hall, Cheshire) and in Kent the use of a deep moulded fascia board was common.

In buildings erected before 1550 the timbers are of great size and placed close together (Fig. 41). This was due to an abundance of timber before this date. As more oak was required for building and ship construction, and despite restrictions placed on its use as fuel, it was found necessary to space the timbers more widely, providing infilling panels of square proportion (Fig. 42). For these panels, in early work, laths were used as a ground for the daub, but in later work hazel or ash wattle was employed. In either case the panels were covered both sides with the clay daub and finished with a coat of lime plaster and white or colour-wash. Other materials were used, chalk or Colleyweston slate for example. The most important material however, was brick used as 'nogging'. This was due to a shortage of craftsmen daubers and the more general use of brick for domestic work. Very fine examples of timber framed houses can be seen in Cheshire, Hereford and Suffolk.

Ecclesiastical work rarely used framed structures for the entire building, restricting the use to particular elements. Timber was used in great quantities for spires, that at Chesterfield, rising to 70 m (228 ft) with its famous twist, being the tallest. The wooden roofs of later Gothic churches are remarkable examples of timber framed structures, especially the single and double hammerbeam and angel roofs of East Anglia. The equally famous hammerbeam roof to Westminster Hall, carried out by the master carpenter Hugh Herland, is one of the finest examples of mediaeval framed structure. Another is the octagonal lantern over the crossing at Ely Cathedral, erected after the collapse of the central tower in 1322. The lantern, reputedly designed by the cathedral sacrist, Alan of Walsingham, and constructed by Master William Hurley, carpenter, is unique. Octagonal on plan, sixteen oak beams coupled together in pairs reach out from the eight supporting

Fig. 44. Pilgrims Hall, Winchester. Hammerbeam roof, c. 1325

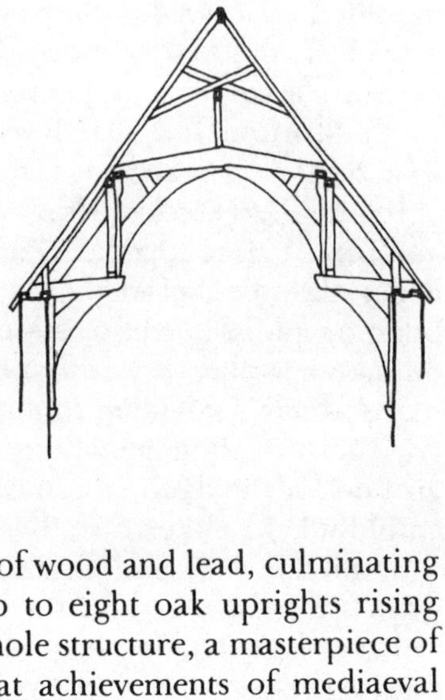

stone piers to support 400 tons of wood and lead, culminating in the lantern itself, framed up to eight oak uprights rising straight for 19 m (63 ft). This whole structure, a masterpiece of engineering, is one of the great achievements of mediaeval craftsmen.

The use of timber as a structural frame has continued in many areas where it is a cheap local product. This is apparent in the New England or Cape Cod style of the Eastern freeboard of the United States, where 18th and 19th cent. timber framed houses of great elegance have perpetuated a tradition to the present day. Now, the trends in the structural use of timber are towards laminates such as plywood. New adhesives, both animal derivatives such as casein and synthetic urea-formaldehyde, enable thin sheets of wood to be bonded together in alternating layers to overcome the natural weakness of timber in the direction perpendicular to the grain, and also to eliminate shrinkage. Numbers of short pieces of timber can now be assembled to form large structural members. The waste, such as sawdust and shavings, is also converted into wallboards with both structural and decorative finishes.

Other materials and combinations of materials have developed to further the technique of the structural frame. Stone, because of inherent defects in its composition, has generally

Fig. 45.   Oil factory, Brisgane, Algeria

been disregarded as a suitable material except in one or two original structures. As with so many building techniques, the best examples are Roman. Most Roman buildings designed for economic activity had a commercial function and there are some surviving examples which can be called factories. The most impressive ruins are those of the oil factories in North Africa, which between the 2nd and 10th cents AD maintained a large population on the production of olive oil extracted in these mills.

At the Brisgane factory the ruin appears as a great cage, the bars of which are formed from enormous blocks of limestone.

The constructional technique was remarkable, comprising small rubble reinforced by large stone trusses set vertically, on top of which rested a horizontal course, again formed from outsize blocks of stone. The corners were formed in a similar way. The arcade arches supported stone beams so arranged that they did not rest on the keystones. These beams were either cut so that their ends rested on headstones on either side of the keystones, or on a single stone beam equal in length to the diameter of the arch. This beam was raised a few inches above the arch by blocks resting on the supporting pillars and embedded in grooves cut in the upper surface of the headstone. Relying on the solidity of the framework, the original mortar has weathered away leaving the structural frame in its original state. This system was first used by the Carthaginians and the technique was borrowed by the Romans during the Punic Wars. Examples may also be found at Pompeii.

The need for large manufacturing units in the 18th and 19th cents, coupled with site restrictions imposed by the necessity to erect factories and mills in narrow valleys, next to either their source of power or supplies of water for manufacturing purposes, necessitated multistorey construction. Where space permitted, it was more economical to erect buildings of several bays with heavy beams and wooden floors supported on rows of vertical supports. Cast iron pillars or storey posts were commonly used, superseding timber due to their smaller size. These pillars are common in warehouses and mills in the North of England. They were also well adapted for use in shop fronts, due to their small size, and many can be seen today supporting the upper storey of older shops in our towns and cities.

Timber beams which were exposed to severe loading were often trussed with iron struts or rods. The ironwork either placed the whole of the timber into tension or, by taking the whole of the tension, placed the timber in compression, thus allowing heavy loads to be carried by relatively small scantlings.

The transition from composite structures to the cast iron frame was accelerated by the erection in Hyde Park in 1851 of the Crystal Palace, designed by Joseph Paxton. This unique structure, incorporating glazing and constructional patents owned by Paxton, led to an immediate interest in this structural form in England. Many of the finest surviving examples of cast iron framed structures are to be found in the Admiralty Dockyards. Perhaps the best is the boathouse at Sheerness, built in 1861, and the earlier smithy of 1858. Another fine example is a storehouse at Portsmouth, erected in 1879. The frank external expression of the frame with boldly expressed infilling of glazed and solid panels is very close to modern ideas.

Efforts were continually being made during the 19th cent. to evolve structural systems to replace the traditional construction of floor and roof. 'Fireproof' floors were provided, constructed of iron girders set 1·8 m (6 ft) apart to support a series of brick arches finished with either a brick or plastered floor. Terraces and flat roofs were constructed with two or more

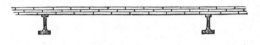

Fig. 46.   Plain tiles and cast iron beams

courses of flat tiles set in cement and laid breaking joint supported at short intervals by cast iron bearers, a system much used in London. The roofs over the Houses of Parliament were designed by Charles Barry of cast iron, the main components being of flat bar, with rolled tee sections forming the principal and common rafters and completed with cast iron struts and purlins. The roofs were covered with galvanised cast iron sheets, lapped at the joints, a great novelty for the middle of the 19th cent.

The first multistorey buildings to be wholly steel framed throughout were the Manhattan Building (1890) and the Rand McNally Building, both in Chicago. In the first, even the party walls were carried by the steel frame, in this case on beams cantilevered beyond their supporting columns. From these beginnings spread the techniques of modern steel framed construction.

The development of reinforced concrete owes much to two Frenchmen, Joseph Monier (1823–1960) and Francois Hennibique (1843–1924). The reinforcement of concrete with steel rods has enlarged the scope of engineers in developing new structures. In addition, the ability to mould the material to the desired form has led to the creation of a greater variety of aesthetic experiences. The idea of prestressed concrete was conceived in 1904 by another Frenchman, Eugène Freysinnet. By replacing ordinary mild steel with high tensile steel wires or rods and stretching or tensioning these prior to pouring, the concrete was brought into initial compression when the tension on the reinforcement was released, sufficient to prevent it cracking when the load was subsequently applied. Although the Freysinnet system was widely used in the 1930s, it was not until a shortage of steel after the Second World War that it was brought into extensive use in England for precast structural members.

The use of lightweight concrete has assumed importance in recent years in reducing the weight of modern multistorey framed structures. Concrete of this type is produced either by the use of lightweight aggregates formed from waste industrial products, such as slag and fuel ash, or by aeration. Lightweight aggregates are produced by various methods, foamed slag by treating the material in a molten state with water. Aerated concrete is usually made by the addition of aluminium powder and lime, which reacting together produce a gas which produces voids in the mass of concrete.

PART TWO

Historical Development of Building Materials and Components

CHAPTER SIX

Stone and Flint

Stone of many varieties has been used for building purposes. Each of the main classifications – igneous (e.g. granite), sedimentary (e.g. sandstones and limestones) and metamorphic (e.g. marble and slate) – has been used in the areas where it occurs, individual characteristics being utilised to their best advantage.

Igneous rocks are generally harder to work than sedimentary, but they can be prepared to fine limits and take a high degree of polish. These are generally used for structures of a monumental character or for the lining to walls or floors. Sedimentary rocks were deposited in successive layers in water and can thus be readily split along more or less well defined planes or natural beds. Their texture is often even and they can be easily worked or carved. Limestones tend to be of finer texture than sandstones and are generally referred to as 'freestone'. Some limestones contain more shell fragments than others and are coarser grained, e.g. Clipsham stone. Metamorphic rocks are found in widely separated areas; apart from slate, which is mainly used as a roofing and lining material, the best known example is marble. Many varieties are found and have been used for building purposes from the first millenium BC.

Where stone was suitable and man had the tools and tech-

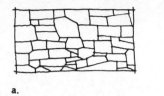

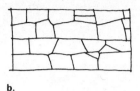

a.

b.

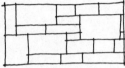

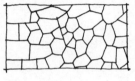

c.

d.

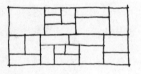

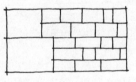

e.

f.

Fig. 47. Masonry
a. random rubble
c. snecked rubble
e. coursed dressed rubble

b. coursed random rubble
d. polygonal random rubble
f. coursed rubble

niques to split it out from its natural bed and work it, walls could be built with close jointing to ensure considerable stability. The method used by the Egyptians to quarry stone was first to cut grooves along the four sides of the block to be detached, then drive in wood wedges, which when saturated with water expanded and detached the block from its bed. The surface of the stone was prepared by pounding it with diorite mauls and squaring up by chisel. Drilling and cutting stone by the use of an abrasive in conjunction with a soft metal blade may well have been practised by the Egyptians, it certainly was by the Romans. Sand was fed to toothless saw blades, hand operated or driven by water power to cut hard stones. Saws with teeth were provided for softer material.

The earliest stone buildings are found in Egypt where an abundance of good building stone, and the technical ability to produce copper and bronze tools for working and preparing it, were coincidental. Stone was, however, reserved exclusively for

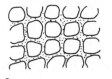

a.

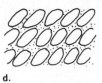

b.

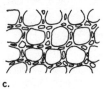

c.

d.

Fig. 48.   Flint walling

a. knapped flint
c. knapped flint with galleted
   joints

b. squared flint
d. diagonally bedded cobbles

civic and religious structures, due no doubt to its cost. Early masonry had uniform courses which averaged about 300 mm (12 in) in height, the joints flush pointed in gypsum plaster. Both limestone and sandstone were used, but in no great quantity until the 18th Dynasty (1500 BC). The most remarkable structures erected in stone were the pyramids. The step pyramid at Sakkara was built with small roughly dressed blocks of brown limestone, cased with a finely worked limestone. The Great Pyramid, the Pyramids of Chufu, Chafra, and the Pyramid of Menkara at Gizeh (c. 2600–2500 BC) were built with a core of local fossiliferous limestone blocks bedded in gypsum mortar, faced with closely jointed blocks of a finer grained limestone. The finest and most impressive of Egyptian temples was that of Amen–Re at Karnak (1400–1200 BC) built of a high quality and carefully bedded sandstone.

The art of fine building in stone was absorbed by the Minoan/Mycenean culture which spread throughout the Eastern Mediterranean. To this phase belong the 'beehive' tomb of Agamemnon (*c.* 1450 BC), generally known as the Treasury of Atreus, and the Lion Gate at Mycenae. The opening incorporating post and lintol constructional elements was relieved by corbel stones.

The stonework of the Hellenic period (700–146 BC) was megalithic in character, finely worked and close-jointed. The Greeks showed great versatility in their masonry construction,

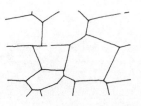

Fig. 49. Polygonal curvilinear masonry.
Temple of Apollo, Delphi. 600 BC

fitting blocks of stone together with great precision. Early walling at Tiryns (1400 BC) comprised large and irregular stones roughly trimmed and piled together, the gaps wedged with smaller stones and bedded in clay. This masonry is known as 'Cyclopean' and developed into a style of walling constructed of polygonal blocks cut with straight facets, a good example being the terrace of the Temple of Apollo at Delphi. Masonry tended to develop into regular coursing, despite some recurrence of polygonal work for decorative purposes. Later work was carried out utilising courses of even height (isodomic) properly bonded and accurately cut with very fine joints. The individual stones forming the South wall of the Propylaea on the Acropolis at Athens show projecting marble blocks or corbels which were used to lift them into position. These blocks would have been trimmed off flush, but the work was interrupted by the outbreak of the Peloponesian War (431 BC) and never finished. A variety in which alternate courses were of even height provided for a proportion of the thin blocks to pass through the full thickness of the wall to form bonding stones. A feature which had its origin in Egypt, and was copied in Mycenean and Hellenistic buildings, was the

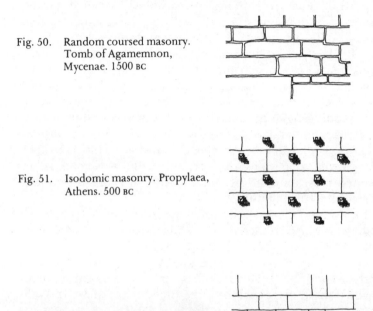

Fig. 50.   Random coursed masonry.
Tomb of Agamemnon,
Mycenae. 1500 BC

Fig. 51.   Isodomic masonry. Propylaea,
Athens. 500 BC

Fig. 52.   Pseudisodomic masonry.
Delphi. 159 BC

occasional use of the slanting joint, first seen in the Temple of Amen-Re and later in the city wall of Messene (450 BC), the gymnasium at Delphi, and the retaining wall of the theatre at Delos (275 BC).

Stone masonry in Italy progressed in a similar manner to other areas of the Eastern Mediterranean. Cyclopean and polygonal walling in the limestone areas of central Italy developed into an irregular rectangular masonry, blocks varying in size but with carefully formed and closely fitting joints. These blocks, which often measured 600 × 1200 mm (2 × 4 ft), sometimes held together with iron cramps set in lead, were laid in horizontal courses, a type of walling called 'opus quadratum' by the Romans.

The Romans had abundant local supplies of stone for their needs but obtained supplies of marble and other decorative stones, for lining floors and walls, from distant quarries. Roman walls depended on the weight and size of the blocks for stability but during the 2nd cent. BC a revolution in constructional techniques producing a mortar which, by gaining considerable strength on hardening, made it possible to erect walls with a core of concrete to carry the load, faced for appearance with small stones (opus incertum). This type of walling became by 100 BC the standard form of construction, and the development of small, carefully cut wedge shaped stone blocks, usually of volcanic tufa (opus reticulatum), with stone quoins, remained in use until the 1st cent. AD. After this date brick quoins became the rule in this type of work.

In Britain, the earliest stone structures were religious in character. The chambered tombs or dolmens of the Megalithic period (e.g. Kits Coty House near Aylesford in Kent) are similar in idea to the pyramids of Egypt, differing only in size and execution. The closely knit social and political influence enabled the transportation of the Blue Stones of South Wales to Stonehenge and the erection of large capstones, some weighing 40 tons, on the dolmens. The possibility of communication between the cultures of the Mediterranean and of the Megalithic builders in Britain may explain the earliest known examples of indigenous masons' work in the mortice and tenon joints of the trilithons at Stonehenge. In the same period dwellings with dry stone walls, sometimes with a backing of earth, were formed at Skara Brae in the Orkneys, roughly rectangular on plan and partly roofed by stone corbelling. In the centre of each dwelling was a stone hearth and on either side beds enclosed with stone slabs. Storage recesses and a stone dresser completed the arrangements.

Stone houses of 200 BC–AD 300 survive at Chysauster in

Cornwall, where traces of eight houses form a village community. The existence in the centre of a stone with a central hollow gives credence to the theory that these houses incorporated timber in their roof structure. At Grimspound on Dartmoor, Iron Age huts were constructed by setting a flat retaining slab vertically in the ground and backing this up with a rubble wall and turves. A stone lintol was provided over the entrance and again, near the centre of the floor, was a stone similar to that found at Chysauster.

Methods of stone quarrying and stone working have changed little in the course of time; picks, crowbars and

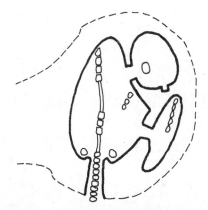

Fig. 53.   Courtyard house at Chysauster, Cornwall. Plan (See also Fig. 3)

wedges are still used for sedimentary rocks. Igneous rocks can only be hewn or blasted out with black powder. Stone is utilised to its best advantage if it is laid on its natural bed, or in such a way that this is set at right angles to the direction of the thrust or load. Where this simple rule is not followed, serious lamination of the stone will almost certainly result.

About 250 English churches contain fragments of Saxon stonework, and where timber was scarce it is reasonable to assume stone was also used for rubble walls of domestic buildings. None of these early structures have, however, survived. The earliest domestic structures now in existence date from the 12th cent. Good examples are the Jew's House at Lincoln and the Manor House at Boothby Pagnell. Several dozen stone houses survive from the 13th and 14th cents, geographically widely separated. In the main, up to the 15th cent. stone was used in England only for ecclesiastical work, for castles and for bridges.

With growing prosperity in the 15th cent., resulting largely from the wool trade, stone became the natural material for both medium-sized and large houses. The Tudor monarchs provided strong government and internal stability, wealth

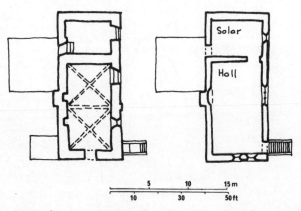

Fig. 54.   Manor house at Boothby Pagnell. *c.* 1200. Plan of ground and first floors

increased and with it a desire for greater domestic comfort. Draughty half-timbered houses were either rebuilt or cased in stone. The mediaeval open hearth gave way to elaborate chimneys and fireplaces of brick or stone, and glass, which in 1500 was rare, had become a common necessity 100 years later. More privacy was required and bedrooms were introduced, often by providing a floor across the old communal hall. After the Dissolution of the Monasteries impetus was given to stone building by the presence of abbeys and priories whose only practical use was as stone quarries, although a few were converted into private houses (e.g. Lacock Abbey). Under Elizabeth I and James I new quarries were opened up and by the end of the 17th cent. locally quarried stone was the usual building material, even for cottages. On the limestone belt, towns like Stamford were wholly built of ashlar. Outside stone-producing areas this material was only used for exceptionally important buildings; while Christopher Wren used Portland stone for his city churches and St Paul's, Kensington Palace is almost wholly of brick.

Although bricks of high quality were readily available in the 18th cent., stone facing was almost universally employed and, for monumental public buildings, considered essential. The reason was social, and while few could emulate the marble and alabaster splendours of Holkham in Norfolk, many aspired to columns faced with scagliola. Smooth-faced limestone or sandstone ashlar was used as a facing on a brick backing, and where cost was prohibitive stone effects were produced by the use of rendering, stucco or rusticated brickwork. Similar facings were provided to timber framed houses, especially in South East England. Stucco gained ground and was almost universally employed up to about 1850. Stone for domestic work has never recovered from this period, even in the Cotswolds and Yorkshire, where fine traditions of native craftsmanship still occasionally produce fine houses in the local vernacular.

An ideal building stone embodies good weathering ability, agreeable colour, texture, and a consistency suitable for the production of a fine face and crisp carving. Very few enjoy all these qualities. The majority of old stone buildings are of rubble, constructed of pieces of stone roughly squared and of varying sizes, arranged without order or direction and brought to course every 300–450 mm (12–18 in) in height. Most stone country farmhouses and cottages are built in this way.

Carefully worked masonry was used for major buildings and an ashlar face on a rubble core was common practice after the 14th cent. By the end of the Tudor period ashlar was familiar in domestic work. This finish was used almost exclusively for all stone public buildings erected from the time of Inigo Jones (c. 1630) until the end of the Regency period (c. 1820). While the rough shaping of the stone block was always carried out in the quarry to reduce the cost of transportation, much of the final finishing was executed on site. Freestone was exclusively used in Gothic and Tudor work for door and window dressings, tracery, string courses and gable copings, and in Renaissance buildings for linings, architraves, keystones, quoins and columns. The worked stones of the vaults, arches and columns of mediaeval ecclesiastical structures illustrate best the perfection of design and craftsmanship in stone. New methods of working stone by powered machinery has cut the cost of natural stone production and with an increasing interest there is a possibility that stone will be used in greater quantity for facing work to framed buildings in the future. Hand craftsmanship, because of its prohibitive cost, will, however, be restricted to the renovation of existing stone structures.

Materials other than granite, limestone and sandstone can well be classified under the generic term of stone. Chalk has been widely used in England, mainly for specialised work. It is easily quarried and worked, and probably the finest example of its use is in the Lady Chapel at Ely Cathedral. The best known quarry was at Tottenhoe in the Chiltern Hills near Dunstable and here was produced stone for Windsor Castle and Woburn Abbey as well as for the lightweight infilling for the vaults of Westminster Abbey. Chalk was used for dressings to brick private houses by Sir Edwin Lutyens but its use has now virtually ceased. Ragstone has been quarried around Maidstone in Kent for about 2000 years, when first the Romans shipped the stone down the River Medway to build the city wall of Londinium. The White Tower of the Tower of London is also constructed of ragstone with Caen stone dressings. The great Jacobean house of Knole near Sevenoaks is wholly constructed of this material. Flint is more commonly used in East Anglia and South East England than in any other area in the world. The material is almost pure silica and was obtained in Neolithic times for tools and weapons from depths of up to 12 m (40 ft) in the chalk areas from such well known mining complexes as Cissbury in Sussex and Grimes Graves in Norfolk. The best flints for building purposes are water worn, but pebbles and cobbles from beaches were also used in considerable quantities. The Romans used flint extensively both for the cores and facings of their walls. Lacing courses of tiles and bricks gave strength and stability and also levelled up the walls for the next lift. Fine examples can be seen at Burgh in Suffolk and Pevensey and Porchester in Sussex. Flint was used extensively in East Anglia and South East England in Romanesque and mediaeval building. This produced such interesting work as the round towers, a feature of many East Anglian churches, and the exquisite knapped square flint and freestone flushwork of which the gatehouse of St Osyth's Priory in Essex is a fine

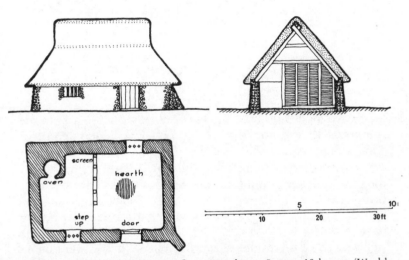

Fig. 55.   Flint and thatch house from Hangleton, Sussex. 13th cent. (Weald and Downland Open Air Museum, Singleton, Sussex)

example. Cobble walls are common in many areas, usually formed with kidney shaped pebbles projecting from the plane of bedding mortar, regularly set in graded rows or set up on end inclining to left or right. Quoins are usually of brick. In some coastal areas the cobble walls are whitewashed or tarred, imparting a vigorous nautical effect. Regency work at Brighton provides many examples of work in this genre. The flint mason's craft is still strong in parts of Sussex, and flint facings to buildings and walls are commonly provided for work in the local vernacular. (See Fig. 48, p. 63.)

CHAPTER SEVEN

Bricks and Brickwork

Men have made and used bricks for at least 8000 years. In the beginning they found that the most convenient method of constructing earth walls was by pressing mud or clay into conveniently sized lumps and, after air or sun drying, to lay them in walls set in mud mortar. Similar bricks have been excavated in Jericho, dating from around 6000 BC, cigar-shaped, 200–250 mm (8–10 in) long × 76 mm (3 in) square, laid more or less in courses. Similar mud bricks are found in many civilisations of the Middle East. By 3000 BC bricks were being made by hand in a mould with a peculiar hog-backed profile, 200–250 mm (8–10 in) long × 150–180 mm (6–7 in) wide and 38–60 mm ($1\frac{1}{2}$–$2\frac{1}{2}$ in) in thickness, often laid in herringbone fashion, with horizontal courses to even up and stabilise the wall (Fig. 56). Bricks of a similar type were used extensively in Babylon and Ancient Egypt, and can be seen in early structures in Spain, Argentina, the Indian reserves of the United States and elsewhere. The Spanish word 'adobe' (signifying an unburnt brick dried in the sun) is familiar.

In the Middle East after 2000 BC, mud bricks were usually cased with burnt brick set in bitumen. In the ziggurat at Aqar Quf near Baghdad, 1400 BC, every sixth course of brickwork had a layer of reed matting running through the mass, to

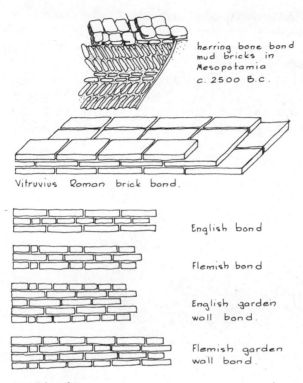

herring bone bond
mud bricks in
Mesopotamia
c. 2500 B.C.

Vitruvius Roman brick bond.

English bond

Flemish bond

English garden
wall bond.

Flemish garden
wall bond.

Fig. 56.   Brick bonds

facilitate drying out and even out shrinkage. From the 9th to the 6th cents BC the Babylonians and Assyrians made patterned bricks and wall tiles with coloured glazes. Fine examples such as the Frieze of Archers at Susa, 500 BC, have survived. This technique died out after the conquests of Alexander the Great and was not revived.

Many kinds of clay material are suitable for making bricks. The material may be pure plastic clay formed from the decay of igneous rock or shale, or it may be alluvium incorporating sand or silt. Colour is due to iron oxide impurities in the clay and conditions of firing. Most clays burn to a red colour when fired at about 1000°C. The presence of vegetable matter in the clay will produce, on firing, a black core.

Bricks are burnt in clamps or specially constructed kilns. Both methods are of great antiquity and have changed little during the passage of time. Clamp burning needs a foundation, usually consisting of a layer of burnt brick to provide a level site and stop rising damp. Channels are provided in the foundation to form flues and are filled with fuel. Green or raw bricks with fuel packed between are so stacked and spaced that fire can penetrate the whole mass. Burnt bricks and mud are

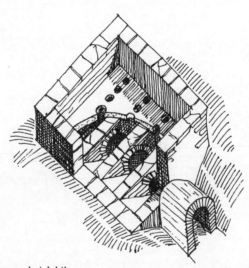

Fig. 57.   Roman brick kiln

laid over the top of the clamp to protect the contents from the weather and reduce heat loss. The clamp is set on fire and allowed to burn out. An early example of a kiln for burning bricks which dates from 3000 BC was found at Khafaje; it is circular in form with four flues below the oven floor, similar in construction to the updraught kiln used by the Romans over 2000 years later.

Fig. 57 is a sketch of a Roman brick kiln based on a 2nd cent. example excavated near St Albans. The structure was of brick and tile bonded with clay and was built below natural ground level to enable the structure to withstand better the stresses of firing, and to reduce the heat loss. The kiln was built on a windward slope and a fire tunnel provided to increase the draught. The oven floor was built of clay and tile fragments with vent holes to draw hot gases up from the flues below.

Roman bricks produced in kilns of this type were of excellent quality. The Roman writer Vitruvius has left us careful instructions for their manufacture and use. He directs that bricks should be made from a white and chalky or a red clay,

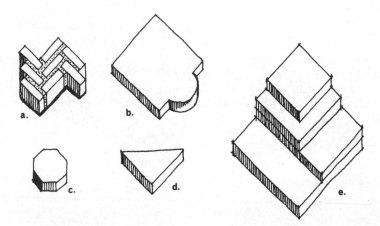

Fig. 58.   Roman bricks
a. paviors
c. brick for column
e. bricks for piers and
   walling

b. moulded brick for half column
d. facing brick

or of a coarse grained gravelly clay, not one which incorporates sand or pebbles. Bricks should be made in the Spring or Autumn and ideally at least two years before required. He describes a method of bonding bricks in a wall which was used by the Greeks (see Fig. 56) which by using half bricks on one face and a course of bricks on the other, bedded to line on each face, bonded by alternate courses laid to break joint, produced a wall with both strength and an attractive appearance on both sides.

Typical Roman bricks were broad and flat and varied considerably in size. A typical Roman brick measured 300 × 150 × 30 mm (12 × 6 × 1¾ in) but bricks 450 × 300 mm (18 × 12 in) are not uncommon and hexagonal and moulded bricks were produced in great numbers (see Fig. 58). The thinness of these bricks enabled them to be very well burnt, the reason for their durability. Mortar joints in early Roman brick walls were between 19 and 25 mm (¾ and 1 in) in thickness but as time passed less care was taken with the work, and towards the end of the Empire mortar joints averaging 38 mm (1½ in), equal to the thickness of the brick, were not uncommon.

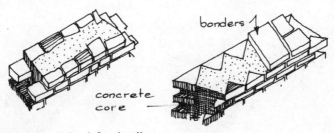

Fig. 59.   Roman brick faced walls

After the fall of the Roman Empire, Byzantine churches of the 6th–7th cents AD incorporated much brickwork in their construction. The walls and vaults of St Sophia in Constantinople (AD 532–537) are of brick set in very thick beds of mortar. The ribbed dome constructed in brick and slightly over 30 m (100 ft) in diameter rests upon spherical pendentives supported by stone arches and piers on a square plan. In spite of adventurous structures of this type, the use of brick in building virtually ceased in Europe and Britain due to the very unsettled social conditions which then prevailed. It is improbable, however, that the craft of brickmaking was ever entirely lost, and in the more peaceful conditions which developed in

Carolingian times (*c.* AD 800) the art once more began to spread across Europe towards the outlying countries.

As we have seen, the Romans used large flat bricks of varying dimensions and these are found in great quantities in Britain. There is no definite evidence that the Saxons made either bricks or tiles, but they did make use of Roman bricks in their own buildings. Ealdred, Abbot of St Albans, demolished the ruins of Verulamium about AD 900, setting aside the whole bricks for the building of his new abbey church. The West front and nave arcades of the priory church of St Botolph at Colchester is also an excellent example of reused Roman bricks and tiles in the 12th cent. A few buildings dated around 1200 survive in East Anglia constructed of or containing local contemporary bricks, but before 1325 most bricks used in Britain were imported from the Low Countries. It is recorded

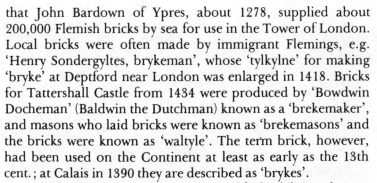

Fig. 60.   Saxon window incorporating Roman bricks

that John Bardown of Ypres, about 1278, supplied about 200,000 Flemish bricks by sea for use in the Tower of London. Local bricks were often made by immigrant Flemings, e.g. 'Henry Sondergyltes, brykeman', whose 'tylkylne' for making 'bryke' at Deptford near London was enlarged in 1418. Bricks for Tattershall Castle from 1434 were produced by 'Bowdwin Docheman' (Baldwin the Dutchman) known as a 'brekemaker', and masons who laid bricks were known as 'brekemasons' and the bricks were known as 'waltyle'. The term brick, however, had been used on the Continent at least as early as the 13th cent.; at Calais in 1390 they are described as 'brykes'.

Mediaeval bricks were burnt with wood fuel in kilns or clamps. From the early 13th cent. sizes varied from district to district, 250–380 (10–15 in) long×125–190 mm (5–7½ in) wide× 45–80 mm (3¼ in) thick. Flemish bricks were smaller, nearer present day sizes, being 200–240 mm (8–9¾ in) long×95–120 mm (3¾–4¾ in) wide×45 mm (1¾ in) thick. It is interesting to note that these latter sizes were almost universally used until 1784 when the tax of that date imposed on bricks produced an increase in the size, as large bricks paid the same duty as small. When in 1803 the duty was doubled on bricks over 2458 cm³ (150 in³) in volume, the size settled down to approximately the present dimensions, being that most suitable for handling by the bricklayer. This duty was ultimately repealed in 1850.

In mediaeval times many bricks were used for paving. In 1368 Master Henry Yevele, who combined his position as Master of the King's Masons with a profitable sideline as a builder's merchant, supplied 1000 'tyles called valthill' for the pavement of the wardrobe at Eltham Palace. Brick was also used for the construction of annealing hearths for glass-making. Excavations for foundations to form footings for walls were often filled with 'bryke battes'. Fireplaces, which came into general use in England in the 15th cent., flues and chimneys, were almost invariably constructed of brick.

It was in the 15th cent. that the craft of bricklaying began to emerge separately from that of the mason. Although in 1442 at Eton they were known as 'brikeleggers', twenty years later at Beverly they were known as 'tylewallers' and in 1505 at Colley-weston 'brykehewers'. Their tools do not appear to have changed much from mediaeval days, and numerous references to the purchase of trowels and packthread for 'lynes' can be found in the accounts of the period.

Good examples of late 14th and 15th cent. brick building are the Gatehouse at Thornton Abbey, Herstmonceux Castle, Tattershall Castle and Queen's College, Cambridge. A possible reason for the increased use of brick in the later period by court architects was the great economy in skilled labour that brick made possible. The warm, pleasant colour of brickwork and the variations made possible by diaper work and the incorporation of stone for detail features was also popular. In the 16th cent. special details were made from a refined clay, known as terra cotta, and there was much use of Franco-Italian or Flemish-Italian decorative motifs, such as are found at Layer Marney in Essex. Cressingham House in Norfolk has

one elevation entirely faced with enriched terra cotta. Bricklayers exploited octagonal forms and showed great virtuosity
in the design of elaborate chimneys in both cut and specially
moulded bricks (Fig. 61). The crow-stepped gable imported
from Flanders was often used with elaborate cusped corbel
tables. The brickwork was generally laid with a thick mortar
joint to take up inequalities in the brick itself.

In the late 16th cent. bricklayers organised themselves into
the Tylers' and Brickmakers' Company, most of whose members produced and laid their own bricks. Members of the
Company leased land and built houses for investment, continuing from the Tudor period their function of combining
design and execution. They made a determined effort to bring

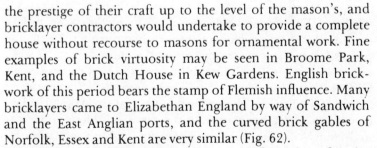

Fig. 61. Tudor decorative chimneys

the prestige of their craft up to the level of the mason's, and
bricklayer contractors would undertake to provide a complete
house without recourse to masons for ornamental work. Fine
examples of brick virtuosity may be seen in Broome Park,
Kent, and the Dutch House in Kew Gardens. English brickwork of this period bears the stamp of Flemish influence. Many
bricklayers came to Elizabethan England by way of Sandwich
and the East Anglian ports, and the curved brick gables of
Norfolk, Essex and Kent are very similar (Fig. 62).

The Act for the Rebuilding of the City of London after the
great fire of 1666 made brick or stone walls compulsory in the
capital. Apart from a few important buildings, brick was
universally employed. Later, many half timbered houses in
East and South-East England were provided with a facade in

brick, a Georgian veneer to a mediaeval structure. Brick as a facing material suffered a setback in the early 19th cent. when Regency architects faced their buildings with a coating of stucco. This material fell out of favour in the later part of the century when the mechanisation of brick production created great savings in time, labour and consequent production costs of bricks. The first patent for a clayworking machine had been granted in the year 1619, but mechanisation did not begin to take the place of manual methods until the middle of the 19th cent. Mechanically shaped bricks came into fashion, and the commercial exploitation of the bed of Oxford clay producing the bricks associated with fletton marked the beginning of the end for the hand-made brick.

Fig. 62.   17th cent. brick gable end

Clamp burning is very little used in brickmaking today, mainly ousted by modern methods of kiln burning. This latter process is carried out using either the intermittent 'Scotch' kiln, the continuous 'Hoffmann' kiln or the continuous tunnel kiln. 'Scotch' kilns, used in small brickyards, have a capacity of about ·50,000 bricks at one firing. They are usually circular, sometimes rectangular, open at the top with holes at intervals in the side walls (Fig. 63). Raw bricks are stacked inside with the spaces filled with small sized coal. When the bricks are dry, the top of the kiln is roofed over with old bricks and clay and the kiln is fired for two to three days. It is then allowed to cool and the bricks are removed. This method of production is similar to that employed by the Romans, and produces a product which is more evenly burnt and contains less waste than that produced by clamp burning.

The 'Hoffmann' continuous kiln, introduced about 1858, is a permanent structure, rectangular on plan, and contains a number of separate chambers, usually fourteen in number, each containing 20,000 to 40,000 bricks (Fig. 64). Each chamber has two permanent walls and an arched roof, dividing walls being taken down and rebuilt as different parts of the kiln

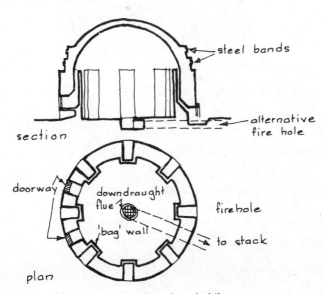

Fig. 63. 'Scotch' or intermittent downdraught kiln

come within the firing zone. One chamber is filled with raw bricks and one emptied of the finished product each day. At the same time another chamber becomes the firing zone. Hot gases from this firing compartment are used to dry the raw bricks. This type of kiln has been considerably developed since its first introduction. The modern continuous tunnel kiln is less common than the 'Hoffmann'. This structure is usually about 60 m (200 ft) long and wide enough to accommodate a brick trolley. Hot air passes down the kiln, drying the bricks before they reach the firing zone in the centre of the tunnel. Here coal dust is inserted from the roof of the kiln to assist in firing.

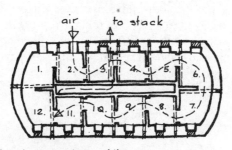

Fig. 64. 'Hoffman' type continuous kiln
1. setting   2. drawing
3–6. cooling   7–8. firing
9–12. drying

In recent years there have been developments in the production of a sand/lime brick, produced by moulding a moist mixture of 90% sand and 10% lime under pressure and curing the resultant product under steam pressure in an autoclave. Production of a concrete brick from a mix of sand and cement has also made some progress since 1945.

CHAPTER EIGHT

Roofing Materials

The word 'thack' was originally applied to all roof coverings. Because the material was usually straw, reed or heather, the word acquired its more limited modern interpretation. Due to its relatively short life, we have little direct knowledge of ancient methods of thatching. It is safe to assume, however, that some form was in general use from very early times. Small clay models of houses dating from about 500 BC suggest a thatched roof. Excavations at Corinth in Greece would indicate that Greek temples from about 750 BC were thatched with straw or some similar material. Thatch was undoubtedly used for roofing Iron Age huts and also by the Romans for their country villas and farmsteads.

Thatch is mentioned in the writings of the Venerable Bede (AD 700) and early writings record its inflammability. We learn that on the occasion of St German's visit to the tomb of St Alban, c. AD 450, a fire broke out and it is noted that the houses burnt in the conflagration were covered with marsh reed. Frequent devastating fires occurred in the Middle Ages; in 1077 London was burnt and in 1161 London, Canterbury, Exeter and Winchester were all destroyed by fire. Despite a coat of whitewash to reduce the risk of ignition by sparks, the danger persisted and the Ordinance of 1212 forbade the use of thatch in London on new work and required that all old thatched roofs should be protected by a coating of lime plaster. Thatch began to fall out of favour for all but the humblest buildings except in those parts of the country where the availability of the raw material and the climate were suitable for its perpetuation. It is interesting to note that thatching was frequently used as a temporary form of roofing for churches in the Middle Ages, and also to provide temporary protection for the tops of walls under construction from frost damage. Walls in Wiltshire and neighbouring counties can still be found with a neat weathering or coping of straw thatch to protect the tops from the weather.

Fig. 65.   Reed thatching
(eaves detail)

In mediaeval times the preparation of straw for thatching by drawing it with a thatching fork to get the straws parallel was usually carried out by women. When this preparatory work had been completed, the thatcher laid the bunches of straw on the laths of the roof, beginning at the eaves and working up towards the ridge. The straw was held in place by plaited straw ropes or rods of hazel. The edges were secured by binding and the ridge formed by overlapping the bunches or by the application of a ridging of clay, turf or sedge.

In the 17th and 18th cents in England thatch was only used for the humblest dwellings. Due to its light weight and original cheapness, thatch was very suitable for roofing buildings formed from cob, clay lump, or pisé de terre and many examples survive in East Anglia and Devon where this form of construction is indigenous. In the 19th cent. thatch was valued for its rural associations. It was used by the architect John Nash, among others, for cottages on country estates. Reed is now considered the best material for thatching, being obtainable in long lengths and with a life of from 70 to 100 years. Rye straw, where available, is a very good substitute, but the original material, heather, has now virtually ceased to be used for this craft. The present condition of the industry is reasonable, there being more work than can be coped with by the

Fig. 66.   Roman roofing tiles

present labour force. The supply of reed is good but straw has become difficult to obtain, as that produced by combining is too damaged for use in roofing.

Tiles for roofing or tilehanging on walls have much in common with bricks, both being baked in a kiln. Often both were baked together, the bricks being placed so as to shield the tiles from the fiercest heat. A better kind of clay is used for tiles, the mix more carefully prepared and consequently the tiles are more uniform in colour and texture.

The earliest forms of roofing tiles comprised two separate units, a wide tile, either square or rectangular and often curved in section and a narrow semi-cylindrical tile usually slightly tapering to one end. The wide tile (tegula) was placed on the roof concave face up and the other (imbrex) placed concave face down over it. This form of tiling is still to be seen in the East. The pattern gave way in time to the broad flat tegula which was the principal covering for the classical buildings of ancient Greece and Rome, and survives in Italy to this day. No examples of the early Greek roofs survive, except a small roof over the underground shrine of Hera at Paestum (*c.* 600 BC). Here the flanged roofing tiles are still in position, but the cover tiles are missing.

For many of their buildings, the Romans used a flat tegula about 430×250 mm (17×10 in) with a semicircular imbrex or cover tile. Sometimes the tile was larger. The tegula was tapered to fit the one below and the imbrices were also tapered to overlap, the joints being mortared or torched to give a weathertight joint. Most tiles produced were stamped with a die to indicate their place of origin, and occasionally the tile bears the mark of the maker. Pear-shaped roofing tiles imitating stone slates or wood shingles were found on the site of a kiln near Ewhurst in Surrey. The ridge of tiled roofs was usually capped with a semicircular tile set in mortar. The ends of the ridge at the gable were often provided with decorative terminals and the free ends of the imbrices at the eaves were filled in with decorative antefixes.

While the Romans used tiles for roofing their buildings in Britain, although in fewer instances than elsewhere in the Empire, there is no direct evidence that this practice continued in Saxon times. The usual roofing material in the Middle Ages was thatch but, as we have seen, the Ordinance of 1212 specifically prohibited the use of this latter material in London due to the fire risk, and one of the alternatives specifically mentioned was tile. It therefore appears probable that the manufacture of clay tiles, copied from oak shingle patterns, became widespread before the introduction of brickmaking in Britain. Archaeological evidence has confirmed the spread of tilemaking to most of Eastern and South Eastern England by the end of the 13th cent. Surnames such as Tyler and Telwright were common by this date.

In early days there was a lack of standardisation in size, and varieties were innumerable. The inconvenience of this state of affairs no doubt prompted the Statute of 1477 which laid down, amongst other things, precise methods for the preparation of the clay and standard dimensions, 265×160×16 mm ($10\frac{1}{2}$×$6\frac{1}{4}$×$\frac{3}{8}$ in). Unfortunately, wood fired kilns were difficult to regulate and local traditions died hard. Kent tiles, for example, for centuries measured 230×150 mm (9×6 in) and in Leicestershire they measured 280×180 mm (11×7 in). The Statute of George I of 1725 confirmed the sizes laid down by the previous enactment but Leicester tiles still held to their own dimensions well into the present century.

Irregularities in clay tiles can be an aesthetic advantage. Tiles were originally hung on riven oak laths which were seldom straight, and secured with small oak pegs driven through holes in the tile and hooked over the laths. Nibs came into general use towards the end of the 19th cent. and are standard practice today.

Good clay tiles are generally cambered in both directions. This allows the tile to lie better and helps to prevent 'creep' by capillary attraction. In addition, the undulations and slight irregularities enhance the appearance of the work. The appearance is also improved by the practical necessity for sprinkling the mould with sand to assist in the removal of the raw clay tile. This results, after fixing, in a pleasing texture. Colour, as with brick, depends in the main on the composition of the clay. The majority of tiles are, however, like bricks, of some shade of red. Special tiles are produced for hips, such as bonnet tiles, and for ridges where the traditional hog back tile is preferable to the half round so common today. The making of swept and laced valley tiles instead of the more common modern purpose-made valley tile has, however, virtually ceased.

The commonest tile used in England was the plain tile, virtually unknown in some countries abroad. Mediterranean countries used the Spanish tile almost exclusively, which provides a covering which is both weatherproof and aesthetically pleasing. The Romans in Britain, as elsewhere, used a combination of flat and round tiles which appears as a compromise between plain and Spanish. A combination of this Roman tiling was developed in the 19th cent. into the patent Roman and Double Roman, originally a speciality of the famous Bridgwater tileries in Somerset.

Pantiles are a traditional English tile form, whose dimensions of 340×240×13 mm (13½×9½×½ in) were decreed by Act of Parliament in the reign of George I. These are, in contrast to plain tiles, of single lap pattern and doubtful weatherproofing quality. Torching was common, and in Norfolk pantiles were often bedded in hair mortar on a bed of reeds. Pantiles were originally imported from Flanders in the 12th cent., and it is generally held that their manufacture in England did not commence until the early 18th cent., and then only in districts which had close commercial ties with Holland and where, in

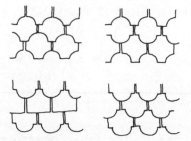

Fig. 67.   Patterns of tile hanging

consequence, the Flemish product was well known. Large numbers of these tiles were used in London in early Georgian times until ousted by the influx of Welsh slates in the reign of George III.

Tilehanging first appeared in South East England towards the end of the 17th cent., to protect the underlying structure from the weather. Although spreading as far afield as Wiltshire and Buckinghamshire, tilehanging has always remained a speciality of Kent, Sussex and Surrey.

The backing to tile-hanging may well be a brick or timber framed structure. Tiles were either nailed or fixed with oak, hazel or willow pins, and sometimes the tiles were bedded and jointed in mortar. The traditional colour for these tiles is a bright red and their shape is often varied to produce scalloped, fishtail, or hammerhead varieties.

Tiles made to imitate bricks were an invention of the Georgian period. It was usual to fix these with nails to battens in a similar way to tilehanging. In Sussex these tiles were known as 'mathematical bricks' and most examples which have survived date from the period between 1784 and 1851 when the brick tax was in force, this form of tiling being exempt. The best examples are to be seen in Lewes in Sussex, where it was the local practice to use only headers. A variation was the black glazed facing tile of Regency days, the best example being in the Royal Crescent at Brighton. In recent years there has been a renewal of interest in this type of tile, especially for use as a cladding for lightweight prefabricated structures.

While the clay tiles are still manufactured and used today in great quantities, concrete tiles which were first produced about 1840 in Southern Bavaria have now made great inroads into the market and bid fair to capture the bulk within the next few years. Early concrete tiles suffered from physical disadvantages such as porosity and loss of colour, but these defects have now been overcome. The main problem, which cannot be easily overcome in manufacture, is their lifeless appearance due to the complete regularity and mechanical appearance of the product.

The Romans used many stone tiles for roofing their buildings, each hung by a single peg from the supporting batten. Various types of stone were used, limestone from Chedworth in Gloucestershire and sandstone from the Forest of Dean. Stone roofing tiles have continued in use up to the present time, using various sedimentary rocks which split naturally into thin slabs 13–25 mm ($\frac{1}{2}$–1 in) in thickness, which can be trimmed to rectangular shapes and sizes. The quarries at Stonesfield in Oxfordshire once covered an area of almost two square miles and were worked continuously from the 13th cent. until the industry died out in the last century.

In mediaeval records 'tiles' are sometimes stone slates. The most famous quarries were at Colleyweston in Northamptonshire which supplied tiles for many of the Royal Works. In addition the slates were used as beam filling between the studs of timber structures, instead of laths or wattle. A few stone slates from the Costwolds were produced until recently, but the Horsham stone slates from the Wealden Beds of Sussex are now no longer produced. Stone slates or tiles were sorted into various sizes and laid in diminishing courses from eaves to ridge, fixed with oak pins to the supporting battens. Owing to their weight, stone tiles required heavier supporting members than other roof coverings and this probably restricted their use.

Shingles are slices of wood, either cleft or rift sawn, used for covering roofs in much the same manner as slates or tiles. In

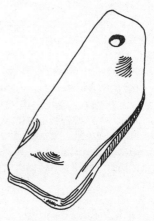

Fig. 68. Stone slate (Gloucester),
304 × 127 × 13 mm
(12 × 5 × ½ in)

Northern and mountainous parts of Central Europe, and North America, they are widely used, and in England from Roman times until well into the Middle Ages. Salisbury Cathedral was originally roofed in oak shingles from the New Forest, and in 1248 Henry III had part of the roof of his manor of Woodstock covered in this material. In 1260, however, orders were given for the shingles to be taken off the roof of the kitchen at Marlborough Castle and replaced by stone tiles. This must have been due to the risk of fire, a risk which was better appreciated in towns, where the use of shingles for roofing had begun to die out, along with thatch, in the 13th cent., in favour of the clay tile. It is likely that the plain tile, so peculiarly English in form, developed from the original indigenous oak shingle.

Shingles continued to be used in country districts. Today they are principally to be seen on church spires in South East England and the Western counties of Hereford, Gloucester and Shropshire. Cleft oak shingles are usually 100–150 mm (4–6 in) broad and 200–300 mm (8–12 in) long, the thick end set towards the eaves. Traditional fixing was by wooden pins to laths, but these have now given way to copper nails. The life of a shingle roof is between 60 and 100 years. Today, the use of oak shingles is rare, imported Canadian cedar having replaced the indigenous oak.

Slates are metamorphic rocks, formed from either clay, shale or fine volcanic ash, either by great heat or continuous pressure, therefore slate can be split easily along parallel planes of natural cleavage. This is of great advantage as it produces thin slabs for covering roofs or for vertical sheathing to walls. Although brittle, slate is extremely hard and close textured, non-porous, resistant to frost and virtually impervious to atmospheric pollution.

The Romans used Swithland slate for roofing in Leicester but after their departure slate as a roofing material fell into disuse until the 12th cent. Then, due to its weight and the difficulties of transport, it was mainly used in Cornwall and Devon, Leicestershire and the Lake District, areas where it was quarried. Large quantities of Cornish slates were, however, shipped by sea after about 1170 to Winchester, Dover and London from quarries in the Charlton and Fowey districts. Slates from Padstow were shipped up the Bristol Channel for the roofing of Glastonbury Abbey. The only quarry now in production in Cornwall is the famous Delabole which mainly produces slate for export.

Swithland slate, quarried in Charnwood Forest near Leicester, could be produced in very large units, blue grey in colour tinged with green, and by the end of the 18th cent. was the characteristic roofing material of the county. After a break from Roman times, quarrying had begun again in the 13th cent. and this continued until economic factors, produced by the cheap Welsh slate, caused the closure of the last quarry in 1887.

The Lake District produces several varieties of slate. That quarried at Furness was sufficiently well known for Christopher Wren to specify its use for the roofs of Chelsea Hospital and Kensington Palace. The slate is produced at quarries close to Kirby in Furness today, and although known as 'blue Westmorland' is in fact a rather dull grey. At Borrowdale the volcanic slate is a very different material, fine in grain and of a beautiful green colour. This material is now very much in demand for cladding.

The ridges to these slate roofs were originally formed of worked freestone, in random lengths of 600–1200 mm (2–4 ft), bedded in lime mortar. With industrialisation, these gave way to purpose-made blue Staffordshire clay ridge tiles in standard lengths.

Fig. 69.   Slate hanging

Welsh slates were in use in Chester from the 14th cent. and by the reign of George III their use was widespread. The material has great practical advantages; it is strong, durable, nonporous, and easily produced in standard sizes. It can be split into very fine laminae, and consequently the weight of this slate for a given coverage has always been less than that of any other slate or tile. Slate can be used on roofs of low pitch and this was a great recommendation to Georgian builders who concealed their roofs behind parapets. The Adam brothers were enthusiastic users, and London rapidly became a slate roofed city.

The first Welsh slate reached London by sea but soon advantage was taken of the new canals to carry slate to inland towns. By the middle of the 19th cent. railways carried the material all over the country.

Due to economic factors the random course method of laying the traditional slate roof was abandoned for monotonous regular courses of 'duchesses', 'countesses' and 'ladies', these being the trade names for the various sizes of slates produced.

The great royal Tudor palace of Nonsuch in Surrey, given by Charles II to his favourite, Barbara, Countess of Castlemaine, on whose instructions it was demolished, was partly finished externally with slates secured to the timber framework. The plaster infilling was left uncovered to expose the elegant and decorative gilded fruit and flowers produced by Italian craftsmen and the first example in England of what we know today as 'pargetting'. The use of slate in this manner is similar to the cladding of concrete structures with slate slabs, fashionable today.

The object of slate hanging was to provide additional protection against bad weather. Sometimes slates were hung on timber framed walls of lath and plaster. Sometimes the slates were hung on battens to a porous stone wall. With brick walls, the slates were generally nailed direct to the joints. Fancy cutting, lozenges, diamonds, fish and beavertails were common, and contrasting colours were also used. Probably the best examples of this work surviving today can be seen in Totnes in Devon.

CHAPTER NINE

Cement, Plaster, Mortar and Concrete

Gypsum plaster was the first manufactured cement. Production probably originated in the Middle East where extensive outcrops of gypsum rock occur. To prepare the finished product, rock gypsum was broken into lumps, stacked in a hollow scooped in the hillside and burnt with wood or charcoal. Gypsum cement was abundantly used in the pyramids at Giza and the tombs at Sakkara in Egypt.

Lime is formed by burning chalk or limestone at about 900°C to convert it into quicklime. For building purposes this is slaked with water to produce hydrated lime, sand and water being added as required to produce a suitable mortar.

As a much higher temperature is required for producing lime than for burning gypsum, it is safe to assume that the production of lime was a more recent development. A method of burning lime, described by Cato over 2000 years ago, is similar to the use of the 'flare' kiln in use up to recent times (see Fig. 70). A rough arch of limestone was formed over a framework or hearth of iron bars and the body of the kiln was filled with lumps of limestone. The fire was kept going until all the material was wholly burnt, taking three to four days. On cooling, the kiln was emptied and then refilled for further use. In Roman times the fuel was timber and often in Roman mortar fragments of charcoal from the fire may be seen. Timber continued to be used as fuel up to the 13th cent. in addition to peat, but due to the vast quantities needed both for burning lime and brick making, grave concern at deforestation was expressed, and coal was introduced as a fuel. As a result of smoke nuisance the use of this new fuel for this purpose was rigorously prohibited in London by the early 14th cent.

The traditional method of slaking lime was to soak the freshly burnt material with an excess of water and then to allow it to flow in a creamy consistency through a metal grating to remove any lumps. The slaked lime was then collected into a container where the excess water could evaporate, leaving lime putty. This was then mixed with sand to form a mortar. If the

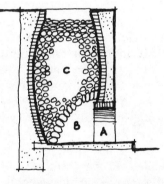

Fig. 70. Flare kiln (intermittent) for production of lime
A. 'eye' or draw hole
B. fuel C. limestone

mortar was to be used for rendering walls and ceilings, cow hair was added as a binder. The incorporation of material known as pozzolana transforms non-hydraulic or semi-hydraulic limes at normal temperatures into stable insoluble compounds. The Romans found that the incorporation of a pozzolanic material known as tufa, a volcanic clay, was excellent for producing hydraulic mortar. In addition to this they used crushed brick or tile which produced a similar effect, a technique used exclusively in Britain due to the absence of local natural pozzolana. These additives produced mortars with similar characteristics and were generally preferred outside Italy. Pozzolanic mortars incorporating crushed tufa were used in places where damp exclusion was important, e.g. for lining baths, tanks and aqueducts, for bonding masonry in waterlogged ground and for sea defence works.

As we will see later, mortars incorporating pozzolana are not found in mediaeval building in England. In the 16th and 17th cents, however, imported Dutch pozzolanas (known as Dutch trass) and lime mixtures were used. Mortars incorporating these materials usually consisted of one part of trass to two parts of slaked lime. The engineer John Smeaton (1724–1792) visited Holland in 1754 to view marine structures and investigated a cement containing equal parts of lime and Dutch trass for use in the works for the new Eddystone lighthouse. He experimented with various limes and found that the lime produced from Blue Lias at Aberthaw in South Wales (which contained some clay) produced the best mortar. He deduced that a good cement could be produced by deliberately mixing and burning together limestone and clay. Eventually the mortar for the Eddystone lighthouse was made from equal portions of Aberthaw lime and a pozzolana from Italy.

Towards the end of the 18th cent. the need arose for a dependable hydraulic cement for the great engineering works then in progress. In 1796 James Parker of Northfleet in Kent found he could make a hydraulic cement by calcining nodules of argillaceous limestone found washed out of the London clay cliffs on the foreshore of the Thames estuary. This cement was called 'Roman cement' and until the middle of the 19th cent. the term 'cement' generally referred to this product. Thomas Telford (1757–1834) used this material in the construction of the Chirk Viaduct on the Ellesmere Canal (1796–1801) and Isambard Brunel (1769–1849) also used it in the construction of the tunnel under the Thames from Wapping to Rotherhithe (c. 1825). This cement was either mixed with equal parts of sand or used neat.

In 1811 James Frost patented a hydraulic cement produced by calcining a mixture of limestone and clay, which he ground together in a wet mill. Improvements to his method were made by raising the calcining temperature and after further development by Joseph Aspden of Leeds the first reliable Portland cement was produced at Swanscombe in Kent by I. C. Johnson in 1845. An early bottle kiln for the production of cement is shown in Fig. 71. A modified kiln called a Dietsch kiln which allowed continuous production was introduced about 1880 (Fig. 72). Rotary kilns were first used at the beginning of the 20th cent., and after considerable technical development are now the universal method for commercial production of cement.

Gypsum plaster, as 'Plaster of Paris', was introduced into England in the 13th cent. when Henry III gave instructions for the material to be used on the walls of Nottingham Castle. He had first seen the material during his visit to Paris in 1254. Plaster of Paris was produced by burning gypsum obtained

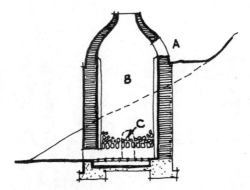

Fig. 71.   Bottle kiln (continuous) for production of cement
A. loading doorway   B. charge   C. unloading doorway

from deposits at Montmartre, Paris. By the end of the 13th cent. a similar material was being produced from gypsum deposits at Purbeck in Dorset.

The use of non-hydraulic lime mortar, that is a mortar without the addition of pozzolanic additives, was almost universal in mediaeval building. This mortar was not so durable as its Roman predecessor. Plaster of Paris is produced by heating rock gypsum to a temperature between 130° and 170°C for about three hours. Modern plasters incorporate this material which, on its own, sets very rapidly when mixed with water. To extend the setting time a retarder such as keratin is incorporated and the material produced is termed hemi-hydrate plaster If, however, the temperature of burning is raised to about 400°C an anhydrous calcium sulphate is formed. This plaster needs an accelerator to increase the rate of drying; modern plasters known as Keene's and Parian are based on this material.

Stucco is a vague term applied to both internal and external renderings of lime, gypsum and cement mixtures. It was used by the Egyptians for levelling up or dubbing out brick walls to provide a smooth and level surface for wall paintings. In Crete

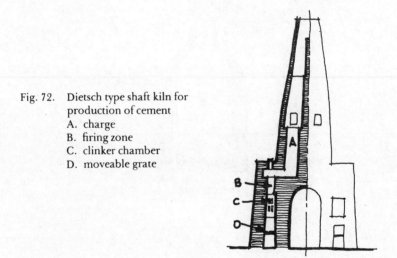

Fig. 72. Dietsch type shaft kiln for
production of cement
A. charge
B. firing zone
C. clinker chamber
D. moveable grate

it was used as a protective coating to rubble walls in Minoan architecture. It was used by the Greeks to protect porous limestone in addition to providing a ground for decoration.

The Romans used stucco extensively and took great care in its preparation. The method of slaking the lime has already been described. The Romans modelled the surface in light relief and their surviving work is of great delicacy. After the collapse of the Roman Empire the art of stucco, although kept alive in the Middle East, remained dormant in Europe. It was revived by the Renaissance and reached its peak in the work of the Italian architect Bernini. The old Italian recipe for stucco comprised a mixture of lime and fine marble dust, with the occasional addition of fine sand and hair.

Italian artists introduced the craft to England in the 16 cent. and Inigo Jones used the material in his work on the Piazza at Covent Garden (1631). The Adam brothers acquired patents for a group of stucco mixtures now known as 'oil mastics' which they called 'Adams Patent Stucco' or 'Adams Cement', and which they used with success in work at Portland Place and Hanover Square, London, in the 18th cent. A similar compound formed from linseed oil boiled with litharge and mixed with china clay and a colouring, known as Dahl's Mastic Cement, was extensively used by John Nash for the external stucco and ornaments of the Regents Park and Carlton House Terraces in London between 1821 and 1833.

Pargetting is the term used to describe ornamental designs in plaster relief. It is nearly always associated with timber-framed buildings, and the designs usually belong to the 17th cent. when the craft was at its height. This craft, virtually unknown outside England, is confined to the period between the 16th and 18th cents.

Although true pargetting is a decorative relief, its beginnings may be found in incised work which often takes the form of a combed decoration. The implements were home made, a pointed stick, or a fine toothed wooden comb. The elevation of the building was often pannelled out and the decoration might consist of simple lozenges or zigzags, or a complicated pattern incorporating a multitude of shapes and designs. This type of decorative work can be seen in Kent at Mereworth near Maidstone, and in widely scattered examples in Suffolk and Essex.

With the advent of Italian plasterers, who worked in England under the Tudor monarchs and, as we have seen, introduced stucco to this country, came true pargetting. Early examples are simple: a set of initials and a date, a Tudor rose. Later more advanced and complicated designs were produced, such as scrolls and swags of fruit, or wreaths of flowers. Good examples of this work can be seen at Dereham in Norfolk and Ipswich in Suffolk.

Scagliola is a coloured gypsum plaster made to imitate marble. The craft originated in Italy and reached England from France in the latter half of the 18th cent. The first known examples were commissioned by the architect Henry Holland from French artists for work in Carlton Place, London, for George III. Scagliola was formed from finely powdered plaster of Paris, mixed with glue or isinglas and applied over an undercoat of lime and hair rendering on a framework of laths. During floating, the colours were incorporated to imitate the particular marble desired. After the material had set the surface was rubbed down with pumice, wiped with a sponge, polished with tripoli powder and charcoal, and finally oiled. Fine examples of this work can be seen at No. 20 Portman Square, London, which was carried out to the designs of the architect Robert Adam (1728–1792).

Mortar is a term loosely used to describe the material used for bedding, jointing, or setting brick or masonry, and comprises a cementatious binding material usually with the addition of a filler. The earliest known use of mortar is plastic mud to cement together the sun-dried bricks used in early structures in Egypt and the Middle East. The great blocks of stone which make up the monumental structures of ancient Egypt were finely squared and worked. A thin layer of gypsum was provided to receive successive courses, not so much for its adhesive qualities but mainly as a lubricant to enable the heavy blocks to be accurately placed. The glazed bricks used in the construction of the city of Babylon (*c.* 2000 BC) were usually bedded in a layer of bitumen, a material which occurred locally in great abundance.

As we have seen before, the Romans took great care in the production and preparation of their building materials, and mortar was no exception. The Roman writer Vitruvius gives precise directions as to the methods for selecting sand for its particular use, fresh pit sand for mortar and river sand for stucco. He also gives advice for checking the amount of dirt in sand and warns against the use of sea sand, which produces a salty efflorescence which spoils the surface. Precise directions are given for the composition of mortars: three parts of pit sand to one part of lime, or two parts of river sand to one part of lime. The addition of a third part of ground brick or tile will, he affirms, give a better mortar when river sand is used. Recommendations on the use of tufa in mortar are also given. The addition of quicklime to mortar as a precaution against freezing in cold weather was also practised by the Romans.

As we have seen, lime/sand mortars were used exclusively in mediaeval building, a mixture which combined relatively little strength and, in coarsely worked masonry, had little life. These mortars hardened slowly, and it was the custom of mediaeval masons to place oyster shells in the joints of the masonry to support the stone and prevent the mortar being squeezed out of the bedding joints under load.

The invention of modern cements produced mortar mixes of great strength. Although the Romans appreciated that strong mixes produced excessive shrinkage, it was not until recently that modern practice reverted to 'lean' mixes with a relatively low cement content, producing greater actual strengths than the much 'richer' mixes formerly in favour.

CHAPTER TEN

Glass and Glazing

Glass has been used for glazing since Roman times, but not on a large scale until the 18th cent. Various other substances have also been used for filling window openings: thin slabs of marble at St Vitale at Ravenna, alabaster, mica, shell, horn and parchment. None of these substances is as effective as glass.

Glass is essentially a compound of silicates of sodium and calcium with small quantities of metallic oxides melted together at a temperature between 1200° and 1500° C and then subjected to controlled cooling. The traditional method of making window glass was by blowing with the aid of a blowing iron, a method still used for bottles and cathode ray tubes today.

Glass-making was well established in Egypt at Alexandria by 1500 BC and from here the craft spread to Palestine, Syria, Greece and Italy. It was not until Roman times that glass was used at all extensively for windows. Panes were generally made from mould blown glass with the use of a blowpipe. This early method produced small but brilliant panes of bottle, slab or Norman glass. After preparing the bubble in the normal way by collecting a blob of glass on the end of the blowpipe, the molten globule of glass was inserted in an iron mould and blown so that the bubble filled it. After cooling, the square bottle sides were cut out for use in panes. In the windows of St Sophia at Constantinople, begun in AD 532, small panes of glass about 180 mm (7 in) square are cemented into a marble lattice. The glass is thick, rather opaque and is probably slab glass.

The use of glass was common in Roman buildings before the destruction of Pompeii in AD 79, for here bronze window frames have been found which would have held glass sheets about 530 × 710 mm (21 × 28 in), and in addition contemporary records mention the use of glass in Roman conservatories.

Glass manufacture spread rapidly throughout the Roman Empire and Byzantine craftsmen became skilled in the production of coloured glass for mosaics. In addition to slab glass, blown sheet or cylinder glass was used by the Romans who made it in a similar manner to the Crown glass of the 17th cent. With cylinder glass, however, the growing bubble was shaped by rotating it in a mould. After reheating and further blowing the bubble was swung to a long cylindrical shape. The end was then opened with an iron rod, and after being placed on a wooden table or 'horse' the blowing iron was detached, the cylinder split lengthwise, opened out and flattened in the annealing oven (see Fig. 73). The production of glass by the cylinder method continued after the collapse of the Empire and the Venerable Bede records that in AD 675 craftsmen came from France to glaze the windows of the church at Monkwearmouth. The glass they used would almost certainly have been produced by this method.

With the contraction of the Eastern Empire, glass manufacturing expanded in Venice and by the 11th cent. the city was the centre of the glass industry, a position it maintained for several hundred years. Much fine glass is still manufactured there today.

The art of staining glass spread to Italy from the Middle East, and again Venice became an important centre for this craft. Glass staining spread to France by the 6th cent. and the art, being fostered by the Benedictine monks, was well established in the 9th cent. In England, stained glass was being extensively used in ecclesiastical work by the end of the 12th cent. and good examples of glass of this period can be seen at Canterbury Cathedral.

White, iron free sand is required for glass making and where this is not available, glass has a greenish or a brownish tint. This colouring is often found in early glass. The addition of manganese dioxide or 'glass-makers soap' corrects this defect.

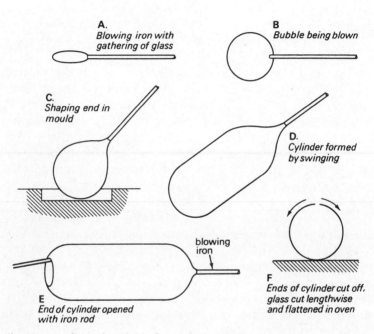

Fig. 73.   Traditional method of manufacturing blown sheet glass

The mediaeval glazier drew his design on his working table, and then drew around the required shape on a piece of glass of the correct colour with a hot rod of metal. This action produced a crack which enabled the unwanted portions of the glass to be broken off. The required details were then painted on the surface and the glass, when heated, had the colour fused onto its surface. The separate pieces were joined with lead strips soldered together, and the whole window was set in rebates in the stone tracery of the opening for which it was designed and steadied with iron saddle bars built into the masonry.

Crown glass appeared in the Middle East about the 4th cent. AD. This type of glass, however, which became common in the 17th cent. is still manufactured today. In this process, after

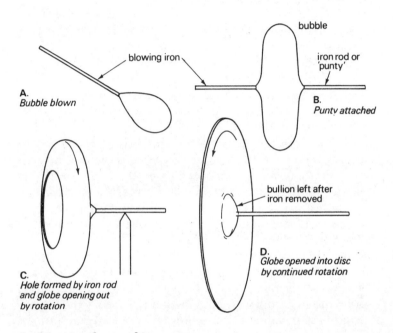

A.
Bubble blown

blowing iron

bubble

iron rod or 'punty'

B.
Punty attached

C.
Hole formed by iron rod and globe opening out by rotation

bullion left after iron removed

D.
Globe opened into disc by continued rotation

Fig. 74. Manufacture of Crown glass

preparing the bubble in the normal way, a solid rod is inserted to form a hole on the opposite side to the blowing iron and by rotating the whole rapidly, combined with further heating, the open shell forms a circular flat sheet or disc. Removal of the rod on cooling leaves the familiar bullion. The disc of glass is often obtained up to 1·20 m (4 ft) in diameter and is normally cut up into flat panes. This glass has a fired or 'flash' finish of exceptional brilliance, produced by the absence of contact by the molten glass with any solid surface until after it has solidified (see Fig. 74).

Since the beginning of the 20th cent. sheet glass has been made by drawing a broad ribbon of glass vertically from a bath of molten material maintained at a uniform temperature. By cooling the edges to solidify the glass, the sheet is kept to a uniform shape and a brilliant fire finish is obtained, the glass being very suitable for window glazing (see Fig. 75).

Cast plate glass, the manufacture of which dates from 1688, was originally made by pouring molten glass onto an iron table and spreading it evenly by means of an iron roller to produce a sheet of uniform thickness. The modern method is to pass a ribbon of glass at 1100°C between water cooled rollers and then through an annealing oven. Both sides of the glass are then ground and polished. This produces a glass of very high quality, but the glass suffers from the expense of its manufacture and the lack of a brilliant finish (see Fig. 76). A method developed within the last few years is to pour the glass onto a surface of molten metal which produces a perfectly flat sheet. By pulling this sheet of glass across the surface and applying heat to both sides a fire finish is imparted. The glass then passes to the annealing chamber where it is allowed to cool slowly. This is called float glass.

Although most churches had filled some, if not all their windows with glass by the 13th cent., few houses had any protection other than horn, blinds or wooden shutters until Tudor times. Glazing was, even then, a rich man's indulgence and glass was so valuable that it was often removed and stored in the owner's absence. Frames were of iron or wood and the panes of glass were usually set in lead cames, which gave the windows of this period their distinctive character. In the 16th cent. glass was arranged diagonally to form a lattice pattern, as at Hardwick Hall, the pane being no more than 125 × 75 mm (5 × 3 in). In the 17th cent. rectangular panes were generally preferred and these were a little larger, about 200 × 150 mm (8 × 6 in). The size was dictated not so much by the dimensions of the glass, as by the pliability of the lead, which could not be relied upon to withstand excessive wind loading.

Sliding sash windows were first used in England at

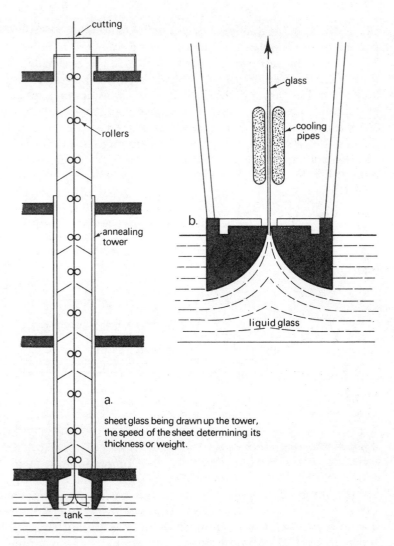

Fig. 75.
a. Method of producing flat drawn sheet glass in a 'Fourcault' tower
b. Detail of drawing trough at base of tower

108

Chatsworth in Derbyshire, between 1676 and 1680, and this
pattern of window was soon being employed in vast numbers.
This must have been due in the main to fashion, especially as
these windows were provided in the contemporary extensions
and alterations to the Royal Palace of Kensington and at
Hampton Court. The sliding sash window had universal popularity and many Jacobean houses were altered by the substitution of the new pattern window for their original casements.
This explains the comparative rarity of survivals of the earlier
type of window. The glass used in the new windows was the
new Crown glass. The windows were divided up into rectangular panes by means of glazing bars, 50 mm (2 in) wide, in
Queen Anne houses, gradually becoming narrower until by

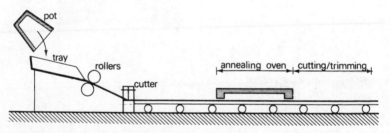

Fig. 76.   Method of producing rolled plate glass

1820 they were no more than 12 mm ($\frac{1}{2}$ in) wide. The early bars
were square and bold in section, later becoming moulded and
more refined. Iron was occasionally used, moulded to resemble wood. The developments of this period were carried
out in the face of a heavy tax on windows and excise duty on
glass, which lasted from its imposition in 1746 until its ultimate
repeal in 1851. By this time glazing bars were fast disappearing
from English windows, plate glass having been introduced
from France in the 18th cent. Cheap cylinder glass enjoyed
considerable success and the glazing of the Crystal Palace in
1851 with 16 oz sheet glass by Chance Brothers ensured its
popularity. New structures were built with uninterrupted
views, their windows being glazed with polished plate or heavy

sheet glass, and there was a wholesale destruction of Georgian glazing bars. In recent years there has been a reversion to original designs, and in many cases glazing bars have been replaced in windows from which they had, regrettably, been missing for many years.

During the present century there have been considerable developments in the production of metal windows from both ferrous and non-ferrous sections. The introduction of the technique of hot dip zinc galvanizing set the seal on the reliability of the steel window, and the production of ranges of standard sizes for domestic buildings brought their price down to the requirements for low cost work. In recent years the production of aluminium and plastic sections has provided a new threat to the traditional wood window.

CHAPTER ELEVEN

Metals and Metalwork

Possibly the earliest metal used by man was copper, obtained by smelting copper pyrites. Production in Egypt dates back to 5000 BC, 2000 years before copper was produced on an appreciable scale in Britain. The sulphur in the ore was partly removed by burning and the resultant calcined ore was placed with alternate layers of charcoal in a furnace. When melting point was reached, the accumulated slag on the surface was removed and the copper ladled into a cavity in the ground to harden. The ingots of metal thus produced varied in size, Roman examples being between 290–380 mm ($11\frac{1}{2}$–15 in) diameter. Most of these Roman ingots carry the official mark.

Copper was not used in building works, except as a constituent of bronze, until mediaeval times when in Holland, Scandinavia and Germany it was extensively used as a roofing material. Copper was not employed at all in England during this period and only very infrequently until about 1800. In the latter part of the 18th cent. copper ores which occurred in the Lake District and Cornwall were exploited and by 1800 Britain was the world's largest producer of the metal. This position was soon overtaken, but during the period when the industry flourished a few important buildings were roofed in copper, a notable example being the North transept of York Minster.

Copper has practical advantages, being the lightest roofing material and thus useful for roofing weakened structures. It also wears well and is impenetrable by beetle. Its drawback is the weathering effects on its colour and the consequent staining of adjacent materials.

Pure copper is very soft, but with the addition of tin a harder material, bronze, is produced. By 3500 BC this alloy was well established for making axes, chisels and saws. Other copper alloys were known, copper and lead for statuary in Ancient Greece and copper and zinc with a small addition of lead to form brass.

One of the outstanding examples of the use of bronze was the roof of the Pantheon in Rome (c. AD 120). This building was once roofed with bronze tiles which were later stripped off and removed to Constantinople. The bronze from the interior of this building was melted down in the 17th cent. to supply the metal required by the architect Bernini for the baldachino over the altar in St Peter's, Rome. The doors for Henry VII chapel at Westminster Abbey, designed and made by the Italian Torrigiano, are also notable examples of work in this material. Bronze was much used for statuary both in Italy by men such as Benvenuto Cellini and by Peter Vischer in Nuremberg, in the 15th and 16th cents. Probably the best example of work carried out in this material in England is the effigy of the Black Prince in Canterbury Cathedral.

Between 3000 and 4000 BC man discovered the art of smelting iron on charcoal fires, producing a material similar to wrought iron. By 2000 BC the practice had become well established. By hammering out or forging the metal and quenching it in water, it was found that the metal could be hardened to produce good cutting tools. In the early days wood charcoal was used for smelting, and since the temperature was insufficient to melt the iron, the resultant product was a spongy mass called bloom. In Roman Britain and for many years after the occupation iron ore was mined in the Forest of Dean and the Weald of Kent and Sussex. At the Romano-British industrial site at Bardown near Wadhurst in Kent, founded about AD 150, iron was smelted until about the end of the 2nd cent. It is estimated that during its lifetime 10,000 tons of iron were produced in the furnaces and evidence shows that the metal was for the 'Classis Britannica', or Roman fleet. The furnaces were of simple shaft type, 300–380 mm (12–15 in) diameter, about 900 mm (3 ft) high and built of clay. Their construction was as follows: a rough hollow 900 mm (3 ft) diameter × 300 mm (12 in) deep was excavated in the natural clay; the walls were then built up inside the hole of prepared clay in sections and when this was dry a carefully prepared internal lining of clay slurry was added. This was dried out by means of a charcoal fire, and any cracks made good with a further application of the clay slurry. During the excavation of this site, no evidence of a forced draught installation was discovered. The front opening in the furnace was stopped up with clay and the molten slag tapped out at intervals into slag basins provided in front of the furnaces. These furnaces were used many times, about 0.65 kg (1½ lb) of bloom or raw metal being produced at each charge. It is possible that production was continuous.

The subsequent development and improvement of smelting owes much to the monastic orders who, as early as the 12th cent., carried out the complete production process. Ironworks were established in Yorkshire during that century by the monks of Kirkstall Abbey near Leeds.

In mediaeval times, most English iron continued to come from the Weald or the Forest of Dean and the smelting was carried out by charcoal. This iron is superior in durability to coke and coal smelted iron due to its low sulphur content. Owing to the demands of the industry, woodlands, especially those principally consisting of oak, were greatly depleted and from the time of Elizabeth I the industry in these areas began to decline due to shortage of natural fuel and restrictions on its use. At Coalbrookdale, Abraham Darby developed the use of coke for smelting the iron ores of the carboniferous rocks of Shropshire and in the early 19th cent. South Staffordshire, Derbyshire, Yorkshire and Cumberland became important centres for the production of iron. Until this expansion, iron was expensive and rarely used in the construction of buildings.

Before the 18th cent. iron was all hand wrought. To help keep it free from rust it was either treated with tin, blackened with pitch, painted or varnished. During this period there was a great demand for wrought iron staircase and balcony balustrades, gates and railings. Many of the best preserved examples of the period are in churches, for example Tijou's ironwork in St Paul's Cathedral and Robert Bakewell's work at Derby Cathedral.

Cast iron was originally of importance in gun founding and by the 16th cent. iron had replaced bronze and brass in this connection. Railings in cast iron did not come into use much before the reign of Queen Anne, and the railings surrounding St Paul's, the Senate House at Cambridge and St Martin-in-the-Fields, London, are fine examples of the iron caster's art. The Georgian practice of placing semicircular fanlights

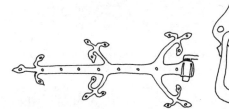

Fig. 77.   Mediaeval hinge   Fig. 78.   Mediaeval door handle
                                                    and lock escutcheon

over doors was carried to its ultimate perfection by the Adam brothers who devised fanlights of great delicacy for mass production in cast iron.

In the Regency period iron became a major decorative element in balconies, verandas, balustrades and fanlights. The ironwork was graceful and distinctive and may still be seen in such Regency towns as Cheltenham and Brighton. The designs were taken from pattern books. After 1830 the quality of design and execution declined rapidly, and the advent of factory-produced mass production made ornamental ironwork a triumph of mediocrity.

Iron ore is now converted into iron by means of a blast

furnace plant. The furnace is about 30 m (100 ft) high and 7·5 m (25 ft) diameter and the ore is fed in at the top mixed with a flux such as crushed limestone. Hot air is blown into the base of the furnace to combine with the coke used as fuel to produce carbon monoxide, which, absorbing oxygen from the ore, is drawn off at the top of the furnace. The molten iron is denser than the slag which floats on the top and the separate liquids are drawn off at different levels. Iron is converted into steel by several processes of which the Bessemer convertor and the open hearth process are perhaps the best known.

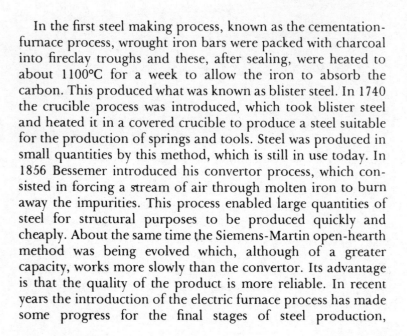

Fig. 79.   Regency wrought iron balcony railings

In the first steel making process, known as the cementation-furnace process, wrought iron bars were packed with charcoal into fireclay troughs and these, after sealing, were heated to about 1100°C for a week to allow the iron to absorb the carbon. This produced what was known as blister steel. In 1740 the crucible process was introduced, which took blister steel and heated it in a covered crucible to produce a steel suitable for the production of springs and tools. Steel was produced in small quantities by this method, which is still in use today. In 1856 Bessemer introduced his convertor process, which consisted in forcing a stream of air through molten iron to burn away the impurities. This process enabled large quantities of steel for structural purposes to be produced quickly and cheaply. About the same time the Siemens-Martin open-hearth method was being evolved which, although of a greater capacity, works more slowly than the convertor. Its advantage is that the quality of the product is more reliable. In recent years the introduction of the electric furnace process has made some progress for the final stages of steel production, especially for special alloy steels where the high production cost by this method can be absorbed.

Before the Industrial Revolution, the only metal in much demand in the building industry was lead. This was used principally for roof covering and glazing. For several centuries before 1850 the bulk of the world's supplies of lead came from England or from Spain. Lead ores occurred generally in association with the older geological formations and in this country were mainly mined in the Northern counties, Derby-

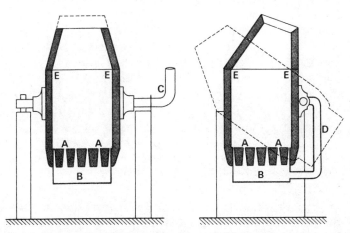

Fig. 80.   Sections through Bessemer convertor   AA tuyeres
           B blast box   C blast   D blast pipe   EE refractory lining

shire and the Mendips of Somerset. These mines date from Roman times. In addition to lead, the ores produced quantities of silver, a valuable secondary product. The last lead mine to operate in this country, the Greenside Mine in the Lake District, closed in 1962.

Until the end of the 17th cent., all lead was cast on a sand bed, about 3 × 1·80 m (10 × 6 ft), in heavy sheets weighing 2·7–4·5 kg (6–10 lb) per ft². The sand left a texture which, while improving its appearance, increased its susceptibility to surface oxidization and consequent decay. Milled lead, which was

Fig. 81.　Siemens–Martin
　　　　　open hearth steel
　　　　　furnace, etc.

rolled very smooth and very thin, was first produced in the middle of the 17th cent., and was used by Sir Christopher Wren at Greenwich Hospital. There it was found to be too light to withstand the weather and consequently its popularity suffered. Most modern lead is milled and of foreign manufacture, being produced in sheets up to 12 m (40 ft) long × 2·50 m (8 ft) wide.

Practically, lead has many useful properties. It is easy to work, melts at a low temperature and can be used cold. It does not break down chemically under normal atmospheric conditions and if correctly used it will last for up to 200 years without serious deterioration. It will, however, when exposed to the weather, wear thin and will then require attention; but it can be repaired, and the old lead salvaged for melting down and re-use. Lead is usually laid over solid boarding and in mediaeval times this was always oak. To avoid reaction from the tannin in the sap, it was usual to lay the lead on a bed of sand. Today, building paper or a layer of bitumen-impregnated sarking felt is used. Lead is best laid in narrow widths about 760 mm (30 in) wide and in lengths not exceeding 2·50 m (8 ft), with lapped horizontal joints and rolled vertical joints. This was mediaeval practice and is still followed in principle today.

Fig. 82.　Lead rainwater head

Lead was, until recent times, the only material which could be used on flat or very low pitched roofs. Many steeply pitched roofs which are now covered with lead must originally have been covered with some other material. The pierced parapets of Perpendicular churches and early Renaissance houses were provided in conjunction with roofs of very flat pitch, universally finished in lead. The pliability of the material was found to be ideal for covering the ogee roofed turrets of Tudor and Jacobean mansions. The limiting factor in the use of lead was its high cost. Set against this was its lightness of weight compared with tiles or slates. As the main consideration was utilitarian, this explains the use of lead in these situations.

Apart from roofing, lead played an important role in glazing, and from Tudor times lead was occasionally used for internal plumbing and generally for rain water goods. Orna-

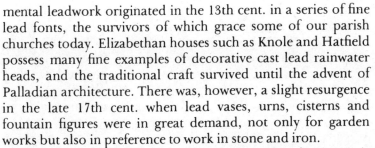

Fig. 83.   Fanlight in cast lead by Robert Adam

mental leadwork originated in the 13th cent. in a series of fine lead fonts, the survivors of which grace some of our parish churches today. Elizabethan houses such as Knole and Hatfield possess many fine examples of decorative cast lead rainwater heads, and the traditional craft survived until the advent of Palladian architecture. There was, however, a slight resurgence in the late 17th cent. when lead vases, urns, cisterns and fountain figures were in great demand, not only for garden works but also in preference to work in stone and iron.

Lead both in sheet form and incorporated in laminates is still used in great quantities today in the building industry, as its long life and excellent properties make it the best material for use in weathering and flashings in building work with any pretention to quality. The use of lead, however, for pipes in internal plumbing has virtually ceased due to competition from other materials introduced in recent years.

CHAPTER TWELVE

Decorative Finishes

The quality of applied decorative finishes in human environment is an indication of the civilising effect of art. At no time in history, except perhaps in the withdrawn atmosphere of monastic life or in our present era, has man by choice accepted natural textures and surfaces as the ultimate in decorative art. He has always striven to impose his own artistic taste on his civilisation, whether it be the cave art of Southern France or the glazed brick pictures of the Assyrian kings. Brick and stone walls were plastered and painted, hardwoods were gilded and decorated with lapis lazuli and other semi-precious stones. Colour was everywhere in the ancient world; the tragedy is that so little has survived the passage of time.

The decorative finishes of Greek and Egyptian buildings were totally dissimilar. The Greeks decorated the smooth marble of their buildings with rich colour, embodying scenes from mythology. None of this work has survived and we are left today with only the descriptions of the classical writers and the unadorned beauty of bare marble walls, sculpture and carving. On the other hand, the Egyptians covered the walls of their temples and mastabas with finely incised figures and illustrations of everyday court life, and while many of these were also coloured, the loss of the pigment has detracted little from the vigour of the sculptor's line. The walls of the great palace of Minos at Knossos were, when excavated, found to have been finished with a smooth plaster decorated with finely executed scenes embodying court life, jugglers and the bulls for which the palace was famed in mythology. The term loosely applied to this form of decoration is 'fresco', used to describe a decorative finish obtained by applying mineral and earth pigments to the damp surface of lime stucco. The colours penetrate the surface and as the stucco dries become an integral part of the work. Work following a similar technique but with the colour applied to gypsum plaster has been found in Egypt, but the true frescos originated in Minos.

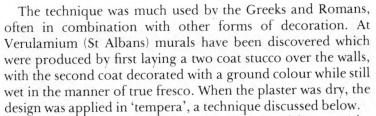

Fig. 84.   Egyptian inscription incised in granite

The technique was much used by the Greeks and Romans, often in combination with other forms of decoration. At Verulamium (St Albans) murals have been discovered which were produced by first laying a two coat stucco over the walls, with the second coat decorated with a ground colour while still wet in the manner of true fresco. When the plaster was dry, the design was applied in 'tempera', a technique discussed below.

The first true modern frescos were produced between the 13th and 16th cents in Italy by artists such as Giotto (1267–1337), Raphael (1483–1520) and Michelangelo (1475–1564). Many magnificent examples of their work have survived. Perhaps the most outstanding examples are to be found in the Basilica of San Francesco at Assisi and in the Sistine Chapel in the Vatican. Giotto found that by adding brickdust to the finishing coat the ground coat was better able to withstand damp, thus following the practice of the Romans in preparing stucco for use in damp situations.

If stucco is first allowed to dry out before the paint is applied the technique is called 'secco'. The work is not so durable and the colours tend to deteriorate beyond recognition. Secco developed after the 16th cent. from fresco because the technique was easier to apply. A third method for decorating a plastered wall was 'tempera'. This is a watercolour process in which water and pigment are mixed with a binder such as egg, glue, honey or milk and applied to a dry surface. Wall paintings executed in this medium are liable to deteriorate if excessive damp is present and while examples in Egypt and other Mediterranean countries have survived, in England this has not been so. Tempera was widely used in mediaeval times, and the practice of whitewashing church interiors in the immediate post-Reformation period preserved many paintings for posterity. During the last 100 years removal of the covering has exposed these to the effects of damp and unskilled preservationists, causing disintegration of the surface. Fine early examples of the technique can be seen at Canterbury Cathedral and Barfreston in Kent, and Hardham in Sussex.

The mediaeval builder and his patron did not hold with the modern view that unadorned rubble walling and bare oak are beautiful in themselves. For the walls, gypsum plaster provided a smooth white surface which, if it became discoloured, could be brightened up with a coat of white or colour wash. Whitewash was usually made up by mixing a bushel of chalk dust in four gallons of size, and it was common to mark out whitewashed walls with red paint to imitate masonry. When a colourwash was applied this might incorporate red or yellow 'oker', and varnish was used for the protection of both wood and metal. Oak beams, panelling and furniture were often decorated with a coat of ochre and varnish, and graining and marbling were practised.

The decoration of churches was almost universal and many examples of mediaeval wall paintings have survived. These range from the crude efforts of village craftsmen to masterpieces which can be found in the churches already mentioned, in addition to Westminster Abbey and Chichester Cathedral.

Royal patronage was not wanting: Edward I employed Master Walter of Durham to execute many paintings between 1260 and 1300 and traces of his work survive on the back of the Coronation Chair in Westminster Abbey. Heraldic bearings were a common form of decoration; in 1240 Henry III ordered his Great Chamber in the Tower of London to be whitewashed, with new shutters provided, and painted with the Royal Arms. Gilding was lavishly employed in buildings of any importance, and the gold leaf employed was expensive. In 1352, gold leaf cost 25p for 100 leaves.

Colours used in mediaeval paints were prepared from natural materials. The best quality azure blue was made from lapis lazuli, but this was expensive and generally cobalt blue was used. In 1290 28g (1 oz) of pure azure cost 8p. For green, verdigris was used, and the price remained fairly constant at about 5p for 454g (1 lb). Red ochre is historically the oldest of all paints and at ½p for 454g was in general use. A bright red was obtained from red lead. White lead was always in common use, and black was usually prepared from charcoal, lamp black or soot. Paint was, of course, not obtained in tins or even ready mixed. Before use the paint had to be ground and made up. As early as 1239 oil was used as a medium for paint, usually a fine quality linseed oil, specially prepared. Consequently it was expensive, between 8p and 10p for 4·5 l (1 gal). While squirrel tails were early used for brushes, badger hair was recognised as being superior. Pig bristles were also used in the making of brushes but, as might be expected, they were usually employed for coarse work such as whitewashing.

After the grim severity of domestic life in the Middle Ages, when every man's house was in reality his castle, the emergence of the strong government of the Tudor monarchs, which provided internal peace to the people of England, was reflected in an upsurge of new domestic construction. Not only were houses for the first time designed for comfort and convenience, but they were also provided with features which were not wholly utilitarian. Panelling, which had become fashionable towards the end of the 15th cent., reached its greatest heights in Elizabethan and Jacobean houses. Unfortunately at the same time the softening influences of domestic life required a more feminine finish. Consequently the acquisition and adornment of walls, especially those of the long gallery which was a feature of these houses, with tapestries was universal. These were often hung three or four deep, and in the great halls which still survived in these new houses friezes of painted and modelled plasterwork depicting hunting scenes or scenes of rural life were provided. One of the finest examples may be seen at Hardwick Hall in Derbyshire. The ceilings of these houses were often finished in stucco with grape or acorn decorated panels, such as may be seen at Speke Hall near Liverpool. Hanging plaster pendants were also popular and good examples are to be seen at Gilling Castle in Yorkshire. A later decorative motif,

Fig. 85. Typical Adam style plaster decoration: Etruscan Wedgwood

universal in Jacobean work, was plaster strapwork, usually found on ceilings, often on walls. None of this work can compare however, with the exquisite work of the 17th and 18th cent. plasterers. Much of the early work was carried out by Italians but in time their English counterparts came to surpass their work in excellence, culminating in delicate and fastidious work carried out for Robert Adam by the Derbyshire plasterer Joseph Rose at Harewood House near Leeds. This work was only matched by woodcarving, carried out by masters such as Grinling Gibbons who enriched doorcases and pediments with a wealth of intricate detail. Much of this carved work was later reproduced in plaster, losing very little in the transition except its texture and deep undercutting. This work was nearly always coloured or gilded and many of the medallions were provided with painted panels specially commissioned for the purpose. A fine example of this work are the roundels in the drawing room at Woburn Abbey, the paintings being carried out by the Italian artist Cipriani. Another great artist in this genre was Angelica Kauffmann, a Swiss by birth, who as a foundation member of the Royal Academy painted several ceilings at Burlington House.

It was in the 18th cent. that one of the most important decorative finishes became available to Georgian architects. Wallpaper had long been one of the delights of Chinese civilisation, and, trade with China being chiefly in British hands, these hand painted papers were sent home as presents by merchants and ambassadors. The finest surviving examples to be found in England are of two types, those with and those without human figures. Some contain hundreds of figures variously occupied with the pleasures of Chinese life, but perhaps the most beautiful are the papers painted with birds and flowers. The chief flower is the peony, and the boughs of the flowering trees are hung with Chinese lanterns and song-birds in cages. Excellently preserved examples, many now over 200 years old, may be seen at Moor Park in Hertfordshire, Cobham Hall in Kent and Temple Newsum near Leeds.

The interest these papers aroused contributed to the 'chinoi-series' of the Rococo period of English architecture, to the Chinese furniture and decorations of Claydon in Buckingham-shire and Sir William Chambers's Pagoda in Kew Gardens. Thomas Chippendale, the 18th cent. furniture maker, pub-lished many designs for pieces in this manner in his famous textbook on furniture design, *Director,* published in 1754.

Mosaic consists of small pieces of coloured material set closely in a bed of mortar to form a smooth and patterned surface. Probably the earliest examples date from 4000 BC at Sumeria, where pieces of stone and fired clay cones were pressed into mud walls to provide a decorative finish. At Ur, palmwood columns were covered with a thick coat of bitumen and encrusted with tesserae of red and black stone and mother of pearl. The Egyptians followed a similar technique using

E

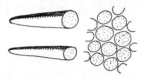

Fig. 86.   Mosaic cones of coloured clay

coloured glass, amongst other materials, inset in tile or stone. The technique of embedding small stones and pebbles in a bed of mortar became highly developed in Hellenistic times and the use of different coloured pebbles allowed designs of great beauty to be prepared. A much later development are 'pitched' floors, once a feature of Welsh farmhouses, where again decorative patterns were formed with small pebbles. The transition from the use of pebbles to that of small cubes of stone was gradual after about 350 BC, and appears to have developed in Alexandria, whence it spread rapidly through the Near East and the Roman provinces.

Many fine examples of Roman mosaics have survived. Various techniques were employed to suit the particular designs being prepared. 'Opus tesselatum' used small cubes of stone or ceramic to produce simple geometrical patterns. 'Opus vermiculatum' used coloured material in an irregular fashion to produce pictorial effects, and 'opus sectile' used marble and coloured material carefully cut to shape to fit the contours of the design. In Britain the usual method of laying the floor was to lay a bed of lime and crushed tile mortar over the concrete base and press the tesserae into a slurry coat of hydrated lime, brick dust and water. Marbles and stones of many kinds were used, sometimes glass or gilded tesserae were employed. The depth of the pieces varied from 25 mm (1 in)

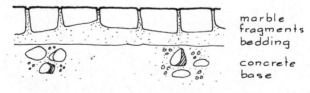

marble
fragments
bedding

concrete
base

Fig. 87.   Section through Roman mosaic pavement

in Rome to less than 13 mm ($\frac{1}{2}$ in) in the provinces, where the cost of material was much greater.

After the adoption of Christianity as the official Roman religion, mosaics were used mainly for the decoration of walls, illustrating religious scenes and figures. The tesserae were manufactured of glass, coloured by metallic oxides, with a great use of gilt. Gold leaf was applied to the surface of the glass and heated, and after flattening and reheating the glass was cut to the required size. Fine examples of this work may be seen at S.Vitale in Ravenna. The use of mosaic in Rome declined towards the 10th cent. but revived somewhat in the development of Byzantine influence, where small fragments of porphyry and precious marble were arranged in bands surrounding discs cut from ancient columns. This development spread to Venice, Southern Italy and Greece.

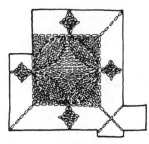

Fig. 88.   Welsh pitched floor

After many centuries the art of producing mosaic and tesserae decoration has been revived in recent decades, and is now a popular method for facing up concrete and similar backing structures, in a way similar to its use over 2000 years ago.

The word tile is strictly applicable to a baked clay roofing unit, the Roman tegula. Today the word has a wider meaning which includes all tiles of a decorative character for walls and floors. Small tiles glazed with brilliant colours were produced in Egypt around 2500 BC, and from 900 BC the Assyrians and Babylonians produced both glazed wall tiles and glazed brick for decorative purposes. The Persians carried on the art, using bricks decorated in relief and coated with tin glazes which, as we have seen before, died out after the conquest of Alexander the Great in the 4th cent. BC.

Neither the Greeks nor the Romans used glazed tiles for decorative purposes. The Romans came nearest to the technique by cutting coloured marble to various shapes and laying them in mortar to form a geometrical pattern. An example of this was excavated at Silchester. The art of tile making was revived in Persia and under Mohammedan influence spread throughout Turkey and as far afield as Spain. In Southern Europe marble and other stones, including the discs cut from columns, were used, as for flooring, but by the 12th cent. ceramic tiles were being produced in Northern France as a

Fig. 89.   English inlaid decorative tile.
13th–15th cents

substitute for these expensive finishes. Tile mosaics and inlaid tiles were introduced to England probably by the Cistercian Order. Good examples may be seen at Byland Abbey in Yorkshire dating from about 1177. These tiles were cut to various shapes to provide geometrical patterns. The glazed surface did not stand up well to abrasion and an improvement was the inlaid tile produced about the same time, good examples being in the Chapter House of Westminster Abbey of about 1255. The Cistercians continued to establish the craft of tilemaking and their abbeys at Chertsey and Great Malvern had their own kilns. The body of the tile was of local red earthenware with the patterns sunk by means of a die and the hollow filled with a white clay. After the Dissolution of the Monasteries tile-making virtually ceased in England, supplies being imported from the Continent. In the 16th cent., Dutch potters from Delft settled in Lambeth and English tin glazed work became known as Delftware.

CHAPTER THIRTEEN

Building Tools and Plant

Tools have existed from the earliest days of human existence: a stick chewed to a point for digging out roots or grubs; a stone with a roughly chipped cutting edge to remove the pelt from a dead beast. Hand axes were an early development of the toolmaker's craft, originally produced from flint or similar hard stone. From this developed, about the time of the last Ice Age, a system of toolmaking in which chips or flakes taken from a core of flint were worked into sophisticated cutting tools. These reached a high degree of technical excellence by about 3500 BC, but the discovery that copper ores could, by smelting, produce a ductile metal, introduced metallurgy to man. The introduction of small quantities of tin to copper, producing the material known to us as bronze, made possible the use of this material for tools requiring durable cutting edges. Weapons, axes, and other objects were produced from moulds after about 3000 BC, and by hammering and annealing provided a sharp and well defined shape. Tools became sophisticated, produced specially for the craft or purposes required. Adzes and axes were used by the Egyptians for roughly shaping timber, saws reduced large scantlings, mortices were cut with hammer and chisel, and sandstone blocks produced smooth finishes. Bow drills were used for the preparation of dowel holes for fixing separate timber members and lathes were used from about 300 BC, very similar to the pole lath used by the furniture industry in England for turning chair legs and stretchers during the 18th and 19th cent.

The introduction of iron to toolmaking brought about a further revolution. Many tools have survived from Roman times and it is interesting to see that most bear a great resemblance to modern patterns. Whereas Roman shovels usually resembled those of the Middle Ages, described later,

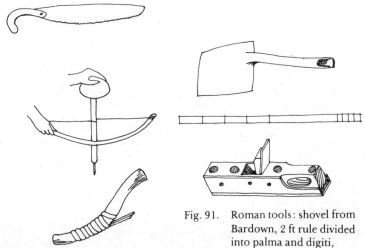

Fig. 91. Roman tools: shovel from Bardown, 2 ft rule divided into palma and digiti, plane from Silchester

Fig. 90. Egyptian saw, drill and adze

one example recently discovered at the Bardown industrial site near Wadhurst in Sussex is made wholly of iron and almost identical to its modern counterpart.

The tools in use in the building industry as a whole have varied little from Roman times until the present day. It should, however, be remembered that iron and steel were once very valuable materials. It was therefore customary for spades and shovels to be made mainly of wood, provided only with an iron shoe. Cutting tools were usually made of iron with a cutting

edge of steel welded on to the blade. Each craft had its own
special tools. Masons used wedges, crows (crowbars), axes,
mauls and 'brocheax' for rough-dressing stone blocks. For
finishing work, masons used the axe and, after about 1170, the
chisel for smoothing the stone to a fine dressed surface.
Labourers used, in addition to shovels, picks for breaking up
the ground as well as mattocks. Masons and bricklayers both
used trowels for setting stone and brick, and saws and 'borers'

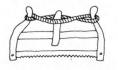

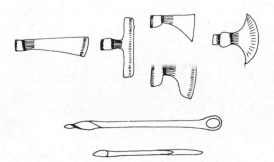

Fig. 92.   Mediaeval frame saw, axe heads, carpenters' augers

for cutting and forming holes. All trades used squares, plumb
rules (levels) and measures.

The most useful tool to the mediaeval carpenter was the axe,
for felling, rough-shaping and splitting timber. Next came the
adze, used for smoothing boards and beams. Two-handed
saws were used for converting logs into boards in the sawpit
and files were used for sharpening their teeth. Awls for prick-
ing out and hammers for driving nails were, together with the
'wimble' or gimlet, the punch and the 'gowge', common to all
carpenters or wrights. Chisels for carving and mortising were

in constant use and the final smoothing of the timber was accomplished by the use of the rough skin of the dogfish, specially dried for this purpose.

Greek classical scholars devoted much time and attention to the study of pulleys. Tackle for lifting heavy weights was obviously in great demand in the construction of large buildings, and the primitive methods of pit and lever employed by megalithic builders were not suitable for the elegant structures of Hellenistic culture. The Romans made use of Greek inven-

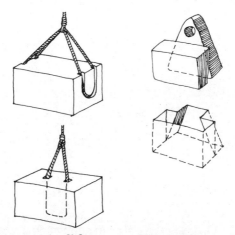

Fig. 93.   Greek methods of lifting stones. 750 BC

tions: gears, pulleys, the screw and the lever. The Roman crane was essentially a sheer leg tripod, the load being raised by compound pulleys motivated by enormous tread-wheels, or by a capstan. Both Greeks and Romans used several devices for lifting individual stones: horseshoe shaped mortices in the ends of the block to take a loop of rope, wedged eyelets and projecting blocks, which were trimmed off when the face of the block was finished after setting.

Cranes and other lifting devices were in common use in mediaeval times for building purposes. A machine often used to lift lighter objects was a 'verne', comprising an upright post carrying on its upper end a short horizontal beam with a brass wheel at either end. This type of lifting device has survived to the present day. Larger examples were made on the treadmill principle, such as that which survives in Bell Harry Tower at Canterbury Cathedral, dating from the end of the 15th cent. Cranes were mainly used for unloading ships and were common in ports by the end of the 15th cent., often used for offloading stone brought by water from distant quarries. The

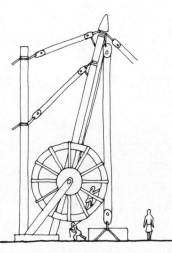

Fig. 94.   Roman crane

'gin' was commonly used on site for lifting loads. This device comprised a rope running over a rope or pulley, to one end of which was fixed a hook, the other end passing round an axle rotated by a wheel. Large blocks of stone were usually raised by means of a 'lewis', slanting mortices being cut in the upper surface of the stone into which the lifting eye was wedged with iron wedges. Another method was the use of two crescent shaped irons with incurving ends, secured to a common ring, the upward strain on which held the irons in position in the mortices provided. Lifting machines of similar patterns to those described continued in use in the building industry until the advent of mechanical power.

Devices for driving piles have been in use from very early days, both piles and a form of sheet piling being used by the Romans. The ram or piledriver used in mediaeval building took the form of sheer legs, supporting a heavy weight by a rope passing over a pulley. The weight, a block of either wood or iron, ran on slides which were greased with tallow to reduce friction. These machines were usually kept in the plant yard and were often hired out for specific works. One, owned by the Wardens of London Bridge, was in 1532 loaned out for work at Westminster. None of these machines has survived but from a description of one constructed in 1329 by Master John de Hurland, carpenter, it appears that the apparatus was controlled by a windlass or crane mechanism which wound up the ram by means of a rope, this being suspended by iron hooks between slides.

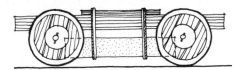

Fig. 95.   Greek trolley for transporting lintols and architraves

While original examples of early forms of transport have decayed and are lost to us for ever, impressions of their shape and the method of their use have survived in clay tablets and incised wall decorations of early civilisations, sledges and carts for moving materials and heavy weights being a common subject.

After about 3500 BC surviving pictures and pottery models show two-wheeled carts and four-wheeled wagons. The wheels were made of solid wood, usually of three shaped planks joined by two cross struts. The axles were usually held in place by straps to enable the wheels to be easily removed, probably for repair. Early carts were attached to the horns of oxen, but the development of the shoulder yoke enabled the wild ass to be domesticated and employed as motive power. Early carts and wagons did not play a large part in long range transport, but were mainly used for short hauls from site to site.

For transporting heavy loads the Egyptians used the sled, occasionally pulled over rollers. Water was poured over the rollers to act as a lubricant, and large teams of men, using papyrus towropes, provided the motive power. It is possible that the sled was moved up a ramp to enable the stone to be placed in position with the aid of levers. It is interesting to note that contemporary illustrations of Egyptian brickmaking show the use of pallets for transporting bricks, considered by many a modern transport technique.

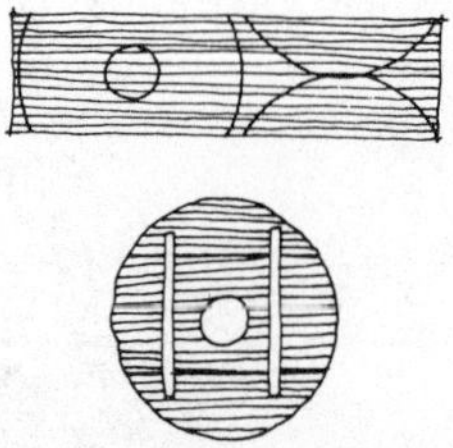

Fig. 96.   Construction of a three piece wheel

Apart from road-building, the Romans did little to further progress in the field of transport with the exception of the substitution of the padded collar for the shoulder yoke. Their carts and wagons were similar to those employed over the previous 3000 years. While Roman ox carts creaked their way over Roman pavé, an unknown carpenter in Denmark designed and made a wheel which, fabricated of wood and bronze, moved on roller bearings. Although this invention died with its maker, it foreshadowed the future of the North European inventive genius which transformed the world in the 18th and 19th cents. Sleds, used for moving large masses of stone, survived into mediaeval times for moving heavy weights. In the main, however, wheeled vehicles were used, drawn by oxen or horses. These wagons were usually hired for the purpose and undertook long journeys, where water transport was not available. Sometimes pack horses were used for small loads. For light work, hand-carts and barrows were commonly used. Often the barrows were constructed with bars to form a barrow for stone; hand barrows were originally a form of stretcher with projecting handles carried by two or more men, usually used to transport large stones for the workshop. Wheelbarrows were also used in great numbers, and barrels and trug baskets used for carrying water and lime. Hods were used for carrying earth up out of ditches and excavations and for carrying sand and mortar. These hods were made of wickerwork, the later wooden hod on a long shaft not coming into use until the 16th cent. Baskets were also in common use for carrying lime and for the removal of rubbish; the term 'to basket out' survives in the specification of building excavation to the present day.

During the whole of this period water was used in preference to land transport for moving heavy loads. Timber, non-existent in Egypt, was imported as deck cargo on frail craft which ventured out from the Nile to bring back cedar from the ports of the Eastern Mediterranean. The Romans built and maintained a large merchant fleet, exporting marbles and worked stone and importing timber, grain, hides and lead from Romano-British ports. The small harbour which served the Roman palace at Fishbourne near Chichester received large quantities of marble and other decorative stones for the execution of the mosaic floors which are such a feature of this building. In mediaeval times, vast quantities of stone were carried by sea from the quarries at Caen in Normandy for the building of many Norman churches in England, and later work of such importance as the Tournai marble fonts, numbers of which have survived in such scattered locations as Lincoln Cathedral and East Meon church in Hampshire. These were transported by boat from the quarries in Belgium to their resting places in this country.

During the 15th to 17th cents, road transport continued to cater for the short haul needs of the building industry, but the development of canals in the 18th cent. provided an impetus for water transport which was irresistible. Roads fell out of use for the transport of heavy materials, and the canal remained supreme until the spread of the railway provided a universal transport system superior to the commercial barge.

For any building work over about 1·20 m (4 ft) in height a scaffold of some kind was necessary. This might only comprise a trestle table, or it might comprise a framework of some complexity such as would be used today. In mediaeval times the platforms were usually of hurdles, planks being too expensive when hand sawing was the only method of timber conversion. Scaffolding was of two kinds: the great scaffold from which the centres supporting the vaults were carried, and where the masons worked, often floored and separated by partitions from wind and weather; and secondly the 'small scaffold', used for access, lighter of construction and incapable of carrying sustained loads. Alder was the timber most favoured for scaffolding and the frame was lashed together with ropes or withies. Often the building employer supplied timber for the scaffolding and centring from his own woods. To reduce expenditure, the highly dangerous practice developed of providing a platform bracketed out from the walls. Sometimes cradles or stages slung on ropes were used, lowered from above. It was not until after 1945 that tubular steel scaffolding replaced a form of timber scaffolding which, with the exception of the substitution of boards for hurdles, had remained essentially unchanged since mediaeval times.

Early units of length were based on measurements applicable to the human body: the cubit – the measurement from the point of the elbow to the tip of the middle finger; the span – from the tip of the little finger to the tip of the thumb of an outstretched hand; the palm – measured across the knuckle; and the foot. In Egypt a cubit equalled seven palms, or 28 finger widths. Elsewhere in the Near East, the cubit was divided into two feet and each foot into three palms. It can be seen that, from early times, standard units of measurement were found to be necessary. Squares and measuring rods were in general use from early times and Egyptian and Roman examples have survived. In mediaeval times, wooden rods or strings were employed for measuring, iron often being used in addition to wood for squares. These instruments were calibrated in a similar way to Imperial measure. Metric measurement, a method completely unrelated to human proportion, was introduced into France after the Revolution at the end of the 18th cent. and owes its popularity to simplicity in use.

Building and civil engineering in the 19th cent. was carried out using plant and tools which had changed very little from mediaeval days. The introduction of the steam engine provided a new and increased source of power which the ingenuity of the inventors of this century turned to good use. The sophisticated plant and machinery in use in the building industry today is based largely on the original thought of these pioneers.

CHAPTER FOURTEEN

Heat and Power

Heat is obtained by the release of energy. The fuel from which this energy is released is today obtained from either deep within the earth's crust, eg coal, oil and natural gas or created by either mechanical means - electricity - or by the absorbtion of natural radiation - solar energy. Originally, timber was the only readily available fuel, burnt on hearths centrally situated on the floor of the building, sparks and smoke finding their way vicariously through vents in the roof. By 1239 in England, charters were granted by Henry III to enable the citizens of Newcastle to dig for coal and in due course the tax on 'sea coal' brought by ship to London contributed to the cost of rebuilding St Paul's Cathedral destroyed in the great fire of 1666.

Coal was the principal fuel of the Industrial Revolution and from its use came the development of the coal gas industry to fill the growing need for chemical distillates and the expansion of gas lighting. With competition from electric light the gas industry turned to heating to maintain economic viability and with the increasing use of natural gas has now captured much of the domestic heating market. Fuel oil, cheap in areas of maximum production, became an important source of energy by the early 1920s and has, despite rising costs, remained so for large installations. Intensive research on solar heating has resulted in the development of economic installations in small buildings situated in areas where satisfactory weather conditions prevail.

Apart from the central log-burning hearth, the earliest heating apparatus was probably the bronze tripod of the Greek household burning carbonised wood or charcoal. A new system developed by the Romans utilised a series of chambers and ducts built beneath the solid floors of their buildings. Called a hypocaust, the construction incorporated a stokehole in the external wall from which smoke and hot gases radiated, discharging from wall flues at eaves level. A fine example survives at the Roman palace at Fishbourne near Chichester, Sussex.

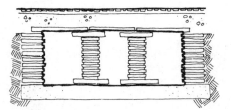

Fig. 97.   Section through Roman hypocaust

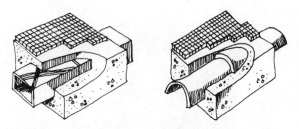

Fig. 98.   Roman heating ducts bedded in concrete floor

After the extinction of Roman influence in Europe, heating methods reverted to primitive central hearths until simple recesses were developed in thick masonry walls connected to the exterior by short inclined flues. A 12th century example survives at Rochester Castle in Kent. More efficient methods of flue

140

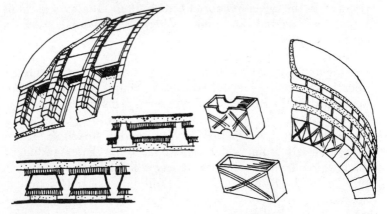

Fig. 99.   Heated vault formed with
           stone ribs

Fig. 100.   Heated vault formed with
            hollow clay ribs

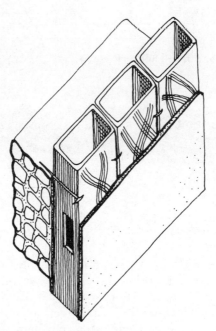

Fig. 101.   Roman flue tiles for wall heating

design improved the draught until by the 15th century
fireplaces assumed the architectural importance they held for
over 400 years, the elaborate decorative chimneys of Tudor
buildings serving to remind us of this fact.

Fig. 102.   Mediaeval smoke-hole in gable of
timbered house

The wasting of natural forests and the growing use of coal
which burnt badly in large open fireplaces led to experiments to
increase efficiency. A notable advance in heating efficiency was
the development in France of the canopy incorporating a
restricted throat or opening into the flue and from this the
principles of modern fire grate design evolved. Enclosed fires or
stoves developed in Europe in the 16th century usually of brick
and placed centrally for efficiency and to promote maximum
warmth. In 1744, the American Benjamin Franklin designed
the wood burning stove which still bears his name. In England,
William Strutt in 1792 produced an iron stove in which air
drawn over a heated surface was distributed by natural
convection and from this idea in 1806 was developed the Belper
stove for use in commercial and industrial premises. Further
work on iron stoves in the United States produced, in 1800, the
first round cast iron stove manufactured in Philadelphia by
Isaac Orr followed in 1833 by the invention of the brass burner
stove for anthracite by Jordan A. Mott.

All these inventions were designed for single compartment
heating. The heating of a number of separate rooms and floors
from a single source developed from the use of hot water in 18th
century France for horticultural heating. Using large bore pipes
and a simple boiler the first commercial installation was for the
new Bank of England premises built in London in 1792. In the
United States, hot water heating was introduced by Robert

Briggs in 1840. Early systems operated by gravity, boilers being placed in basements. With the introduction of motor driven pumps or circulators, boiler location became a matter of convenience rather than simple physics. Steam heating, invented by James Watt in 1784 using waste steam from boiler testing to heat his workshop, was patented by a man called Hoyle from Halifax, Yorkshire, in 1791. One of the earliest commercial installations was for a silk mill in Watford by Thomas Tredgold in 1824. This method of heating was introduced into the United States and was preferred to other systems. Improvements included the 'overhead down feed one pipe system' introduced by John H. Mills in 1877 and Frederick Tudor developed the modulating valve and in 1902 made improvements in the installation known as the 'return line system'.

Warm air heating was popular in the United States by the 1930s. Warm air from a furnace was carried by gravity through metal ducts and introduced into rooms through grilles. To reduce the problems of frictional resistance from the duct walls small motor-driven fans were introduced to move the air through the duct. Return air was collected, filtered and cleaned before being recirculated. With the introduction of a cooling coil, year round air conditioning was provided. Radiant heating, developed from the idea of the Roman hypocaust, originated in England from the work of A. H. Baker in 1908 involving the embedding of heating coils, warm air ducts or electrical resistance panels or cables in floors, walls or ceilings.

District heating was introduced in Lockport, NY, in the United States in 1877, using steam from a central plant to heat a number of separate buildings. The idea became popular as an outlet for waste steam from electricity generating plants. In England the first installation was in Manchester in 1911 and one of the largest constructed for the Pimlico Housing Development in London using excess steam from Battersea Power Station immediately across the River Thames from the site.

Gas as a natural phenomenon was first noticed in a coal mine near Wigan in England by John Clayton in 1664. Experiments were carried out to produce 'inflammable' air by the distillation of coal and these continued until 1784 when Jean Pierre

Minclers first used gas produced by such a method for lighting. The first commercial exponent of this new source of power was William Murdoch, a former employee of James Watt, who first illuminated his home in Redruth, Cornwall, in 1792 using gas produced in large iron retorts and conveyed in metal pipes to the burners. In 1795 Murdoch constructed a small experimental production plant enabling him to carry out his first commercial installation in a Birmingham factory in 1798.

Municipal gas lighting and power distribution commenced in 1813 by the formation of the London and Westminster Gas Company. Originally comprising three production plants employing 15 miles of underground mains, the company provided lighting to the city streets. Expansion was rapid and by 1823 the company had laid over 122 miles of mains within the City of London and were providing over 250 million cubic feet of gas annually. Rapid expansion of the gas industry ensued, gas companies being set up throughout Great Britain regulated by local and general acts of parliament. Each company marketed gas on its own terms and, to ensure fair and reasonable tariffs by the introduction of a unifying method of quality evaluation, in 1920 the Gas Regulation Act made it obligatory for companies to charge for their gas on the basis of declared and attested heating values and supplied at a statutory pressure. On 1 May 1949 the gas industry in Great Britain passed into public ownership.

Early production of commercial gas was carried out in deep brick shafts, this method being abandoned on the introduction of the Siemens process in 1861. This system allowed coal to fall from a hopper to lie on a producer over a step grate. Connected to a furnace, air for gasification was drawn through the fuel by natural chimney draught. Ever growing demand led to the development of producers working under a positive blast of air and steam. Byproduct recovery from producer gas was high, yielding large quantities of ammonia and tar through lymm static washers. New processes, notably the Lurgi process developed in Germany in 1945, used small coal of high ash content generally unsuitable for other processes and at Southend in England producing rich gas by the partial combustion of oil. Today, almost the whole of gas supplied for heating and industrial purposes in Great Britain is obtained either from natural wells situated in the North Sea or by importation in a liquefied state in specially constructed ocean-going tankers.

Electricity, as a physical phenomenon, had been known for a great many years before Humphrey Davy in 1809 attached two charcoal rods to a large galvanic battery and, by moving the rods together, produced a brilliant ray of light. This laboratory experiment had two practical limitations. Firstly, the charcoal rods burnt rapidly away but after much experimentation the French physicist Leon Foucault substituted carbon rods for charcoal and produced a clockwork mechanism which, by moving the rods closer as they burnt away, maintained the arc of light. Secondly, the galvanic battery used by Davy was large and cumbersome and new methods of storing electricity were required.

Batteries may have been developed by the Parthians about 250 BC for gold plating jewellery but in modern times it was Alessandro Volts who, in 1792, observing the difference in potential of metals in contact with an electrolyte, developed the electrochemical series described in a paper delivered in 1800 to the Royal Society. This was the battery used by Davy in his experiments, producing electric current but with no storage capacity. In 1859 Gaston Planté discovered the ability of lead plates to increase electrical currents resulting from the polarisation of the chemical reaction occurring on the surface of the metal immersed in sulphuric acid. This cell, after a short charge, could produce a moderate amount of current. Improvements were devised by Thomas Edison who in 1900 suggested the combination of an alkaline electrolyte containing an iron negative plate and a nickel oxide positive plate producing a practical and efficient storage battery.

The dry cell battery was invented in 1865 by Georges Leclanche and incorporated a slush of sal-ammoniac, zinc and carbon sealed and made watertight in a container. This invention led to the later development of the mercury battery where the prime source of energy is the oxidisation of zinc giving a relatively constant output throughout the life of the battery.

The electric generator, converting mechanical power into

electricity, originated in an experiment conducted by Michael Farraday in 1861 using a copper disc rotating edgeways between the poles of a horseshoe magnet. From this discovery Hippolyte Pixii in 1832 constructed a generator using permanent magnets and coiled wire armatures. Continuous experiments proceeded until in 1856 Ernst Werner Von Seimens developed the shuttle winding and produced the world's first commercial generator. The copper and alloy brushes on early generators caused problems which were solved by the introduction of carbon brushes by C. J. Van Depoele in 1888. All these generators produced direct current (DC). The invention of alternating current (AC) power generation was the joint work of John Gibbs and Lucien Ganiard in Europe and William Stanley in the United States. Originally powered by steam turbines, water wheel generators are now common where water supply at pressure is available and since the 1960s nuclear energy has developed as a source of motive power.

The electric lamp developed from experiments conducted by Nicklolas Hawksbee who, about 1706, discovered that an evacuated glass globe containing sulphur, rotated at speed, produced a faint luminescence when rubbed. Many experiments finally produced the earliest recorded truly incandescent lamp by De la Rue in 1820. The first public installation using this lamp was in Ohio, USA, using lamps produced by Charles F. Bush. The problem of short life filaments was eventually solved by Thomas Edison who, building on work on suitable filaments carried out by Sir Joseph Swan, on 21 October 1879, successfully lit a lamp incorporating a carbonised thread filament. The metal filament lamp was a product of the work of Welsbach and von Bolten culminating in the tungsten filament of Alexander Just and Franz Hanaman of Vienna. The introduction of the coiled coil filament in 1937 using an inert gas to retard filament evaporation is the basis for modern incandescent lamp manufacture. Fluorescent lamps were developed in the 1930s and first introduced commercially in 1938.

The development of the electrical supply network progressed in parallel with the development of electrical generators. By 1878 plans were being prepared in England by St George Lane-Fox to provide electricity to specific customers from a common supply system. Operating on a low voltage DC current, the first successful system came into being in London on 12 January 1862. The use of low voltage DC caused problems but following pioneering work by S. Z. de Ferranti, alternating current was adopted in Great Britain, the first power station being constructed at Deptford. The system spread rapidly throughout the country utilising local supply companies with their own generating plant. With the greatly increased use of electricity after World War 2, very high voltage transmissions needed to maintain power supplies were introduced by the Central Electricity Generating Board after nationalisation, with a 275,000 volt AC supergrid linking centres of power production with the original 132,000 volt grid reverting to feeder lines.

The need for heating, power and lighting in buildings has progressed in parallel with the development of these services. From such opportunities offered to designers, great advances have been made in both environmental control and personal convenience, by the introduction of sophisticated heating and lighting techniques and increasing mobility through the introduction of elevators and escalators and advanced methods of communication in modern buildings.

PART THREE

Building in the Middle Ages

CHAPTER FIFTEEN

Organisation of Mediaeval Building

The lodges were the centres around which the organisation of the craft masons evolved. These were either temporary or permanent, depending on the size and length of the contract. Some cathedrals and the greater monastic houses maintained permanent lodges where a small nucleus of masons and carpenters carried out their work under the control of a master mason. The latter was usually appointed for life and, being continually resident, was required to give his first attention to the buildings of his employers. In 1435 we learn that Master Richard Beke was appointed by the Chapter of Canterbury to sole charge of the masonry of the Cathedral. He was to receive a salary of 20p weekly, a house, clothes, an allowance for fuel and a pension on disability. On the other hand, consultative appointments were sometimes made where the mason received a small retaining fee, being paid a wage when he worked and having freedom to accept other commissions when not engaged in his consultative capacity. This was the type of post to which William Attwoode was appointed by the Chapter of Wells Cathedral in 1490 at a fee of £1·33. Sometimes an employer enticed a good mason from his previous employer as in the case of William Hyndlee, brought from Norwich to York, to whom the Dean and Chapter of York paid £5 for, among other items, defending the suit brought against him in London by his enemies, 'without just cause'!

Master masons were men of high standing, enjoying positions of authority, with a retirement pension in their old age. Their remuneration was commensurate, frequently a salary of over £18 per annum at a time when the possession of land worth £20 per annum compelled a landed proprietor to become a knight. They sometimes received payment in kind, Henry Yevele in 1389 receiving two manors in Kent in lieu of his salary of 5p a day. The master mason always dressed in keeping with the dignity of his office and the terms of his contract often, as we have seen, included the provision of clothes. The chief mason at King's Hall Cambridge at Christmas 1431 received robes to the value of 83p.

In most crafts the term 'master' designated one who, having served his apprenticeship, had set up in business on his own account. Only in the masons' guild (and to some extent in the carpenters') was he almost certainly an employer, responsible for the body of craftsmen under his authority. It was his duty to engage and discharge workmen and, where he was acting as contractor, to pay the men in his employment. An early instance of the uncertainty of commercial life is shown in the contract for the Church of Fotheringay where, in 1434, a clause allowed for the payment of workmen direct by the Clerk of Works if the contracting mason, William Horwood, failed to do so. The cost was to be deducted from future payments to Horwood.

The unusual position of the master craftsman was due to the large number of day rate journeymen. Labour then, as today, was fluid and large numbers of men worked under one master, many of whom would be fully qualified to rank as a master in their own right. The principal qualification was to have served an apprenticeship under a master for a period of at least 7 years. During this period the apprentice received board, lodging, instruction and sometimes pocket money. Profit from his labours went to his master. At the end of the 7 years the apprentice usually served a further period as a journeyman until the opportunity arose for him to act as a master by undertaking contract work. Then, under the regulations of the

London guild, he would require four masons of some standing to guarantee his capability.

Building operations might be carried out under a Clerk of Works by a staff of workmen, or the work let out by contract. In the first instance, master craftsmen were engaged as foremen to deal with the technicalities of their respective crafts. In the formation of a contract, the usual method was to provide a simple statement of the nature of the building, its dimensions and a list of such features as doors, windows, and fireplaces, with a note of the contract sum and the penalty for failing to

Fig. 103.   Drawing in the master builder's workshop

complete the contract. Sometimes the building was specified in great detail, or existing buildings referred to as examples to be copied. Occasionally references are made to future plans or other drawings. Materials were sometimes provided by the contractor, at other times by the employer.

Payment was usually made by providing a sum down in advance and the remainder in instalments either at regular intervals or as the work progressed. The carpenters who constructed the roof of Hartley Wintney Church in 1415 were paid £10 when they started work, £6 when they had framed up the timbers and a further £6 when the roof framework was complete. The contractor was usually required to provide a bond for the proper execution of the work and often produced sureties bound jointly and severally with him. Occasionally bonds were mutually exchanged between employer and contractor.

In addition to the contracting staff of craft workers, most contracts needed a small administrative staff to pay the wages and keep accounts. Payment for parchment and wages of clerks to keep the 'journals' or rolls of daily expenditure are frequently found in the accounts of major projects. In very large contracts clerks were appointed to deal with matters on a specialist basis, one for wages, one for the purchasing, and so on.

The regulations prepared for the employment of craftsmen laid down in minute detail the hours of work. A distinction was drawn between winter and summer, the dividing dates usually being Easter and Michaelmas. The working day was defined from dawn to sunset, but there are instances of work continuing by artificial light. In 1365, 168kg (370 lb) of Paris candles were bought at Windsor for carpenters and masons working at night by the king's orders. There were intervals set aside during the day for rest and refreshment. Timekeeping was usually by means of a bell, but at Chertsey Abbey, in 1538, 2p was expended on an hour-glass to regulate the workmen's hours. On Saturday work usually stopped at 3 p.m.

Holidays usually coincided with saints' days and church festivals. Christmas and Easter provided a brief holiday varying from four days to a week. Events of national importance were sometimes the pretext for an unauthorised holiday. An example was the birth of the future Edward III in 1312 when the workmen at Westminster took a week off. Although the frequency of church festivals gave the workman some rest from his labours it also reduced his earnings, as he was not paid on these occasions. It was the custom, however, for those employed on the Royal Works to receive one day's pay for each two days' holiday.

The number and continuity of royal building projects poses the problem whether any centralised agency was employed for their management. Our present knowledge is not sufficient to answer this question adequately but in view of the difficulty in computing the quantities of materials required for the Royal Works in London – Westminster Abbey, Westminster Palace, and the Tower of London – it was possible that a store or dump was set up to enable the various superintendents to draw material for their needs at any one time. There was certainly a store at Westminster by 1250, where stone was supplied to the king's order for use at Westminster.

Large ecclesiastical projects needed special adminstration. Usually this position was filled by a separate official, rather than the sacrist, who was normally responsible for building repairs and the completed fabric. This official, known as the 'custos fabricae', was usually appointed annually though some are known to have held the position for several years. Little is known of their detailed duties. All we know is that they had to see that the revenues set apart for the fabric were duly collected and to arrange payment for the materials and labour used in the work.

Even less is known of municipal building work, the boroughs being less active and surviving records few. Most guildhalls were of wood and contracts were carried out in one of three ways:
(a) by direct labour and direct purchase of materials
(b) by the direct purchase of materials and the hire of labour through master craftsmen
(c) by lump sum contract to an appointed contractor, usually a master craftsman.

Contract administration in municipal work was usually entrusted to custodians or keepers. In some instances, instead of looking after the work themselves the boroughs contracted the work out to craftsmen. The adminstration of private works was similar to royal and ecclesiastical projects already mentioned. On occasions, however, the owner acted as his own Master of Works and accountant, hiring his own workmen and buying his own materials.

The size of mediaeval buildings is no indication of the size of building operations. Many buildings were erected piecemeal over a very long period of time by a small permanent body of workmen carried by the Chapter establishment. With other projects the building works were carried out in one great concerted effort involving the simultaneous employment of a large number of men. In many of these contracts the situation of the site played a large part in the organisation and two principal patterns were required to deal with these different situations:

(a) the 'ad hoc' organisation specially set up for the erection of a castle or abbey

(b) the continuous organisation needed for the construction of cathedrals and bridges of importance.

The first matter to be settled after the selection of the site was the acquisition and opening up of a quarry for the stone. In some instances stone was purchased from existing quarries and in others stone from Caen in Northern France was imported specially for the purpose. The final selection of the material was left to the master mason. Considerable quantities of timber were used in all mediaeval building work, not only for roof and floors but also for shoring, scaffolding and piling. Most of the smaller scantlings could be readily obtained from local carpenters or dealers, but the method for obtaining large beams was to select suitable trees, negotiate their purchase and then fell and prepare them. Various methods were used to obtain lime for mortar. Sometimes it was purchased ready for use, sometimes the raw materials were supplied to a lime burner who was paid a wage for his labours. Sand was often dug on site but sometimes cartage was paid to bring the material to the site and on occasion labourers were paid to dig and load it onto the builders' carts. Bricks were often made and burnt on the site, being selected as a principal constructional material according to the local availability of brick earth. Otherwise they were purchased from the brickmaker and transported to the site.

The carriage of building materials was an important problem for the mediaeval builder. The cost of land transport was very high and in many cases those responsible for building works organised their own transport. Where transport was hired, teams were usually engaged by the load or by the day. Where water transport was used the usual method of payment was either at a rate per ton or for the employer to hire the ship and crew for an agreed sum for the round trip.

With the difficulties that existed in computing exact quantities for works in hand, inevitable surpluses occurred and these were disposed of in the normal commercial way. In 1433 the Chapter of York Minster sold ashlar to the City of York, and in 1456 to Beverley Minster. There was also a brisk trade in second-hand building materials, this latterly becoming so profitable that, after the Dissolution of the Monasteries, great abbeys such as Vale Royal in Cheshire were so completely removed and utilised for the erection of other buildings that no trace of them today remains above ground.

As we have seen before, the lodge was the organisational headquarters of the workmen. These buildings provided both workshop and living accommodation and were the prime responsibility of the carpenters at the commencement of work. None of these semi-permanent structures have survived but the construction most certainly would always have been of wood or wood and wattle. At Westminster, in 1413, one lodge was roofed with tile and one with reed. The craft tools were kept in the lodge and the building was no doubt heated in cold weather, as we find fire-stones mentioned in the rolls of expenses.

Assistance and attendance was afforded to the craftsmen by the employment of servants and labourers. These latter mixed and carried the mortar and raised the stones onto the scaffolding. Sometimes these men worked in gangs, levelling the ground, moving earth and stone with wheelbarrows and digging and laying foundations. Many of these men may have been recruited or impressed at the beginning of the contract for this particular labour, returning to their agricultural work as soon as it was completed.

It will be seen, therefore, that the mediaeval building organisation was a complex affair, closely controlled and regulated by craft laws and yet enjoying a freedom of expansion and development far greater than permitted today. Many of the regulations governing craft works are part and parcel of building today, and their survival is proof of the qualities of their originators in mediaeval times.

F

CHAPTER SIXTEEN

Mediaeval Craftsmen, Designers and Administrators

In mediaeval times, craftsmen could principally be divided into two crafts, those of masonry and carpentry. Masons included within their ranks several sub-divisions. The commonest term for the craft is 'caementarii' – stonehewers – and in early documents the terms 'mazones' or 'mazoni' may be found. The superior branch of the craft were known as freemasons or workers in freestone, the material used either carved or faced for ashlarwork, as opposed to rubble masonry. Comparisons were made between the freemasons and masons called 'ligiers', i.e. layers or setters of stone. The second class of masons therefore consisted of the setters who placed in position the stones worked by the (free) masons. A third class of mason was the 'hard-hewer' who was employed to get the stone out of the quarry.

Carvers of stone were originally inseparable from masons, the carving of capitals and ornamental work being, throughout the 11th and 12th cents, inseparable from the architecture of the building. Certain men, however, even in this period, emerged as carvers of great aptitude and artistic merit, their work having come down to us in the magnificent Romanesque fonts and tympanums to Norman doorways which embellish many of our smaller parish churches today. Even in the 13th and 14th cents many statues were carved by masons who worked on the mouldings and window traceries. The work tended to become more specialised, and men with a special aptitude and feeling for this work began to emerge from the general ranks of masons. These carvers were referred to as 'imagers'. There was a flourishing school of monumental masons near Corfe in Dorset, who worked the Purbeck marble produced from the local quarries which was in great demand in England during this period. These marblers included the polishers and formed a further sub-division of the craft.

On the other hand, carpenters were more catholic in their activities and, while sometimes using the appendage 'wright', in general carried out all types of woodcraft from felling timber to manufacturing tilepins. The conversion of timber was carried out in saw pits where the sawyers worked in pairs. Joiners were sometimes mentioned in accounts in connection with fittings and furniture, and choir-stalls were often worked by 'karvers'. Connected with but distinct from the carpenters, were tilers, slaters and thatchers. Plumbers and glaziers, smiths and painters were also, in a lesser capacity, connected with the main crafts, and lower in the scale came the plasterers and pargetters and the daubers who made the cottage walls of wattle and daub. The pavior working in stone, marble or tile executed the floors, and from the 15th cent. the 'tylewallers' or bricklayers made their appearance. Without taking into account the mortar men, hod men, barrow men, limeburners and manufacturers of tiles and brick, and the carters and carriers who moved the materials, it is obvious that the mediaeval building industry was a highly developed and complicated organisation, employing a great deal of labour.

As we have seen before, from the beginning of the 13th cent. the building industry was organised into a system of craft guilds. These guilds were local and gave a man no right to work or exercise his craft in any other town. This form of organisation was not suitable for the masons who were constantly on the move from one contract to another. Here, the local permanent guild was replaced by a temporary association based on the lodge or craft workshop where the mason was employed. The reason for the acceptance of the local guild by the carpenters, glaziers and other crafts was because there was in every town, where buildings were constructed mainly of timber, sufficient employment for a good number of local tradesmen. Stone buildings, on the other hand, were rare. In fact, in most towns the only stone building was the church. Thus, unless he enjoyed a permanent post on some structure of size, the mason spent much of his time travelling from place to place in search of employment. A project of any size would more than exhaust the local supply of labour, and thus account for the movement of craftsmen which was such a feature of mediaeval building.

The number of men employed on a contract varied with the degree of urgency of the work. In 1295 the average number of men employed on Caernarvon Castle during June and July was 398, but it is unlikely that private contracts, even for the

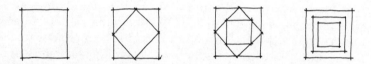

Fig. 104. Mediaeval proportion

wealthiest patrons, employed such numbers. While a castle was of little use unless rapidly constructed, the building of a church might be spread over several generations. To recruit the craftsmen necessary for their works, the English kings impressed their labour. These workmen were enrolled for the King's Works on pain of imprisonment. They were, however, well paid, receiving rather higher than general rates to compensate for the inconvenience. In addition to labour, materials and carriage were also obtained for the Royal Works in this manner.

When large numbers of workmen were provided for a contract, the problem of accommodation and provisioning was considerable. It was usual at the start of the work for the carpenters to be employed on the construction of lodges and dwellings for the masons and other workmen. Where the work was close to a town the men would be billeted on the towns-folk, but accommodation was still necessary. At Dover in 1536, a request was made for tents for dining in bad weather to prevent the workmen going into the town for their meals and thus wasting the king's time. The lodge was the building on which centred the life of the craftsman. Besides being his workshop, it also served as dining room and often as a dormitory. There is nothing to show that modern freemasonry has its roots in mediaeval craft masonry. There was, however, a bond of unwritten tradition, jealously guarded, which with a certain knowledge of geometry invested the masons' guild with a certain air of mystery.

Geometry, a science unknown in Western Europe in the 11th cent., was vital to set out and construct the vast works of Romanesque and Gothic architects. Most of the ancient Greek texts were lost to Western scholars during this period and men were not capable at this time of finding out by themselves anything which could be called geometric science. Some of the knowledge they did possess came through the study of the works of Vitruvius, but mostly this came from Arabic scholar-ship. During the 9th and 10th cents Moslem scholars translated the works of classical antiquity into Arabic, and a school of translators at Toledo in Spain translated these into Latin. By the middle of the 12th cent., Greek scientific culture expressed in the arts of algebra, geometry and trigonometry was freely available to Western European scholars, architects and crafts-men. It cannot be supposed, however, that the architects of this age had a profound knowledge of these matters. Their knowledge must have been, in the main, empirical, and they showed a clear preference for simple proportions, e.g. 2:1, 3:1, and so on. Units of measure varied from one city to another, and architects avoided scaling their plans. They were also vitally interested in proportions which could be trans-posed from plan to actual scale or dimension without using a linear scale.

Professional organisation changed radically between the 12th and 13th cents. At that period, architects began to organ-ise themselves on the basis of the guild system, and to restrict the publication of technical and scientific knowledge acquired

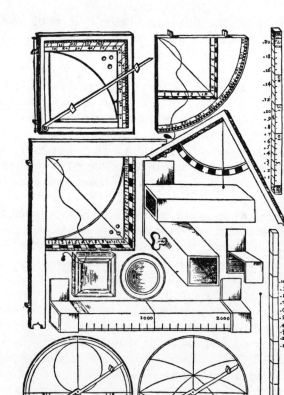

Fig. 105. Mediaeval measuring instruments

during the previous period. Amongst this information was Vitruvius's solution to 'doubling the square'. This was used, not only to determine harmonious proportions for such elements as cloisters or pinnacles, but also to establish plans of certain churches.

By the end of the 13th cent. the architect was fully aware of his worth. His status had changed and he was no longer engaged in the workshop. The word 'magister' preceding the word 'cementarius' frequently indicated an architect in mediaeval records, although the title 'magister operis' is often an indication of a foreman. The Master of Works in England was very often a royal appointment to supervise works in progress.

During the great era of cathedral building, architects do not appear to have made models of wood or plaster of their designs. This means of communication, popular in Romanesque times, did not reappear until the Renaissance. Very few plans have survived from this period. A sketch book by the French architect Villard de Honnecourt, now preserved in the Bibliothèque Nationale in Paris, includes elevations but they are more in the nature of research than original working drawings. Again, very few mediaeval drawings of any sort have survived, possibly because the high cost of parchment prohibited its use as a drawing material and consequently sheets of plaster or whitewashed boards were used. The use of the latter material is common even today for setting out full size details in many cathedral workshops. The best known 13th cent. designs surviving today are the Reims Palimpsest, being a series of drawings showing the design for Strasbourg Cathedral. The document, dating from about 1250, is on parchment which has been reused time and again. One design, that of a large church, is of especial interest. The elevation is bisected by a vertical axis, broad outlines only being shown on the right half, details on the left. Detailed specifications, such as those drawn up in 1284 for the Church of Grey Friars at Provins, complemented the plans. As these plans were, as we have seen, rarely dimensioned, it was usual for the specification to give measurements for the principal elements.

Architects or 'principal masters' received a larger salary than the other workmen, their social and financial status being higher. This was because of their ability to organise and direct a workshop, to draw plans and prepare specifications. Cathedral chapters found themselves in the position of applicants for the services of exceptional men. These latter, being limited in

Fig. 106. Timber cutting

number, could negotiate advantageous contracts, and often refused to restrict their freedom of movement to a single workshop. Sometimes emoluments were negotiated in curious ways. In 1129 Master Raymond, architect of Lugo Cathedral, worried lest currency values dropped during the term of his employment, asked to be paid principally in kind. In addition to six silver marks, he demanded silk, firewood, shoes, salt and candlewax. The advantages of payment in kind were numerous. Architects were often housed free, they were given robes and could be exempted from taxation. They were often given a bonus at the end of the year, and their employment was usually for the year, the life of a workshop or, rarely, for life. Architects were also often engaged to provide expert advice. This was usually in connection with the collapse of a vault or after a fire. In this case the Chapter concerned would often invite several well known architects to survey the damage and draw up a report before carrying out repairs or re-building. These advisers were always properly housed and fed, and received sums of money which compared very favourably to their annual incomes.

However good the designers and craftsmen, some authority was needed to deal with the day-to-day administration of the work and to arrange for finance to be available. In addition, an authority was required to smooth the way and to keep the project moving to the satisfaction of the employer or patron. These administrators were often men of great ability, who could and did rise to high position in their professions, often obtaining preferment as a reward for their services. One of the greatest mediaeval building administrators was William of Wykeham. Although not a professional architect, it was his capacity for superintending building work for Edward III that set his career on a firm foundation. He entered royal service about 1347, and in October 1356 he was appointed surveyor of the works at Windsor. Three years later he became surveyor of the castles of Leeds, Dover and Hadleigh in addition. He entered the church in 1361 and became Bishop of Winchester in 1367. His chantry chapel in this cathedral epitomises his love of beautiful buildings.

CHAPTER SEVENTEEN

Mediaeval Building Methods, Materials and Costs

Most materials used in mediaeval buildings were of natural origin, stone and wood being the most important. These were cut from the quarry or the forest and after weathering or seasoning were prepared and incorporated in their allotted position in the structure. Certain materials, used in relatively small quantities in relation to the total volume of building, were manufactured – lead and tiles for roofing, lime for mortar and daubing, glass and pigments for enriching the design. We have seen how these materials were produced and used in mediaeval structures and how manufacturing techniques expanded to cope with increased demand. No account has yet been given of the costs involved either in the production of natural and manufactured building materials or the installed cost of works. Nor have methods been discussed which were employed in the incorporation of these materials in structures. Some information is now available for the study of this subject due to the preservation of mediaeval account rolls and contracts.

Foundations for mediaeval buildings were usually formed from a well compacted layer of stone, chalk or flint laid in the prepared trench. A binding of strong lime mortar was usually employed and walls and columns were constructed directly off

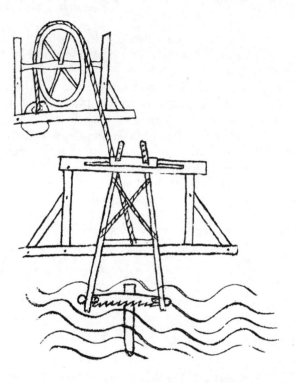

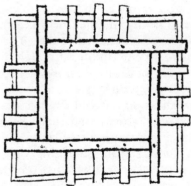

Fig. 107. Cutting a timber pile under water

this levelled base. In wet or marshy districts elm, beech, alder or oak piles were used, driven into the soft ground until a proper set was obtained. At Sandwich in Kent in 1463, elms for piles were purchased at 5p each. To help preserve the timber it was the usual practice to char or scorch it before use. In addition piles were used for the foundations for wharfs and piers, and in the construction of bridges piles constituted the whole of the foundations. When Old Rochester Bridge was demolished, it was found that 10,000 elm piles 6 m (20 ft) long and shod with iron had been used to support the structure. The piles were covered on the sides and top with elm planks which gave support for 200 mm (8in) layer of Kent ragstone which formed the foundation for the masonry.

Foundation stones for buildings of any importance were laid with some ceremony and usually by some person of importance. Stone walls in mediaeval days may be divided into two

Fig. 108.   Masons' marks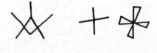

classes – those composed entirely of irregular blocks of stone set haphazardly in mortar, known as rubble walls, and those formed from carefully worked rectangular blocks laid in regular courses and called ashlar. The core of most ashlar faced walls was stone rubble, although the walls to parapets and the thin walls of later clerestorey construction were of solid ashlar masonry. In many cases ashlar was used as an external facing only, the rubble-faced inner surface being plastered. As we have seen before, oyster shells were used in large quantities to even up the courses of ashlar walls and prevent the slow-setting lime mortar from squeezing out of the joints under load. During frosty or wet weather the tops of walls under construction were protected by straw or heather thatch.

Windows of all shapes and sizes were provided in mediaeval buildings. Circular and 'rose' windows were a feature of many cathedral gables (Westminster Abbey) and oriel windows were common (Kenilworth Castle). In those days windows were described as of so many 'lights'. Sometimes the window recess was carried down and finished with a seat. A few references occur to the cost of windows, a window in three lights of stone at Langley in 1367 cost 75p, and at Woodstock a 21·34 m (70 ft) run of freestone for windows cost 1p per ft run.

A notable feature to be found from early times in important dwellings was the fireplace. In 1400, at Eltham Palace, 500 bricks were provided for the construction of three fireplaces; the construction of kitchen fireplaces for castles is an item which frequently occurs in account rolls. Such fire-

Fig. 109.   Tower scaffolding

places would incorporate ovens. While many houses still used the common hearth, fireplaces were being constructed against an external wall with the flue carried up, as we have seen before. Many of these have survived until the present day in the otherwise empty shell keeps of contemporary castles (Rochester Castle).

As a large part of the cost of masonry lay in carriage, it was imperative that local material should be used as far as possible. Water carriage was far cheaper than land transport and this is possibly the reason for the widespread use of Caen stone in the coastal towns of South and South East England. Many large monastic foundations owned and operated their own quarries. Quarrying was a regular trade and the greater part of the stone was produced in rough hewn blocks of convenient size. At Harlech in 1286, two quarrymen were paid £1.25 a hundred for quarrying and cutting two thousand and forty stones approximately 600 × 450 × 300 mm (2 × 1½ × 1 ft). As the size of stone was not usually mentioned in the records, it is probable that a stone averaged 0·028 m³ (1 ft³). Another common measure of stone was the ton, based on the weight of a tun of wine (908 kg) (2000 lb). In 1382 stone for St Paul's was supplied by Master Henry Yevele as 45p per ton.

A large proportion of stone used was for ornamental work—plinths, string courses, battlements, etc. This was often supplied ready worked from the quarries, cut to measure in stock patterns as was standard practice in the 15th cent. When a quarry opened and the overburden was cleared away the top layer of stone, always inferior, was removed for use as rubble and filling. The lower layers of stone were always of better quality, and so long as the quarrymen ensured that the stone was properly seasoned after cutting and dressed to lie in the wall in the same position as in the quarry, all was well. Mediaeval masons were not always careful to observe this latter precaution, sometimes with disastrous results. Initially, the blocks of stone were got out of the quarry by heavy mauls or hammers and steel wedges, and reduced to approximate sizes required for the job by further splitting or sawing. The final tooling of the stone for ashlar was often carried out on site. Masons' marks are frequently found. Each mason had his own mark, and used this throughout his career. Sometimes these marks were cut in the quarry and sometimes on site. As a rule, the deeper cut the mark the older the work. (See Fig. 108.)

Bricks came into use in the 14th cent., mainly in the Eastern counties, due to Flemish influence. Mediaeval bricks were either burnt in kilns using wood as a fuel, or in clamps; a small brickworks making bricks in the mediaeval method fired by oak cordwood was in production at Ashburnham in Sussex until 1969. In 1440, 9500 bricks costing 21p per thousand were provided for work at Shene. By 1505 the cost of bricks had risen to 50p a thousand, although these may have been picked facings. By this period moulded and ornamental brickwork was the fashion, and we learn that in 1536 John Baldwyn was paid 75p for 'hewing' or cutting 5000 'angle, chamfer, skews and rounds'.

Lime and sand were used in large quantities for mediaeval building. Lime was either bought ready burnt or special kilns were constructed near the site for the purpose. Ready made lime was bought by the load or the quarter, or in small

Fig. 110.   Moving a pillar

quantities by the bushel. A load of lime for Hampton Court in 1533 cost 28p, and at Salisbury in 1483 1½p a bushel. Where kilns were constructed their cost is often shown in the accounts; at Winchester in 1222 the sum of £2·50 was expended on this item. While vast quantities of wood were consumed as fuel, where 'seacole' was available this was commonly used. At Purbeck in 1291, twenty loads of coal from Newcastle was purchased for the sum of £4.

After the middle of the 13th cent. plastering came into general use, following its introduction by Henry III at Nottingham Castle. Gypsum was the material used, and early plastering made use of imported material. Plaster of Paris was imported into Southampton, where in 1532 we find that 1 ton for the vaulting of Winchester Cathedral cost 17p. Nottingham was famous in mediaeval times for its alabaster, and the waste and inferior stone from the workshops was burnt for plaster. In 1504 William Upton sold 62 tons for 15p a ton (including carriage) to Colleyweston. Once plastered, the application of decorative finishes to walls and ceilings soon became fashionable. At the Tower of London in 1337, £3·67 was paid for pargetting and whitening the whole of the Great Hall, a very large decorating contract for those days.

By the second quarter of the 12th cent., glass was common in churches. In accounts, coloured glass was always separated from white, the latter in 1240 costing 2p per $0·093 \text{ m}^2$ (1 ft^2), the former being double this price. Old glass was often re-used, due to its value, and the cost of the labour was reckoned at 1p for 304 mm (1 ft) in addition to the cost of material for leading and fixing. Standard designs and illustrations were produced by the larger glassworks and the cost of the glass depended on the degree of skill employed and the work involved. Between 1445 and 1447 John Prudde, the King's Glazier, produced glass with roses and lilies and armourials at 40p, glass with different figures and borders at 50p and glass incorporating the finest figurework as supplied for the Beauchamp Chapel at Warwick for £1 a square metre. Even in those days damage from careless carriers was frequent and glass was consequently packed in cases well provided with hay and straw. Glaziers usually worked in gangs, making the glass, preparing the windows and fixing them. The place of manufacture is rarely mentioned in accounts but it is known that glass was produced at Chiddingfold in 1225 and the Chapter of Salisbury had their own glassworks. Much of the best coloured glass was imported from the Rhineland, Flanders and Normandy.

The importance of masonry construction must not overshadow the use of less permanent materials. The majority of buildings erected in mediaeval times were of humbler construction, walls of turf and sods, earth floors and roofs of poles covered with 'thack' of turf or straw. The use of mud walls or pisé de terre has survived in certain counties to the present time. Wattle and daub, consisting of a row of upright stakes filled in by interweaving hazel or willow rods and faced up on both sides with clay or plaster, finished with a coat of white-wash, was a common walling method from early times. That this method was used for internal partitions can be seen from the payment of 17p for new spars and wattling to a house at Grendon in 1500. Many examples are to be found in dwellings which have survived from this period (The Priest's House, Alfriston, Sussex).

The timber roof is regarded by many as the chief glory of mediaeval craftsmanship. The angel roofs of East Anglia and the hammerbeam roofs of Westminster Hall and Hampton Court are well known examples. These roofs were put together, as were all mediaeval roofs, by jointing and framing and pegging with oak pegs, every effort being made to stabilise the thrusts and exert as little overturning moment to the support-ing walls as possible. Many of these roofs were framed up in the carpenter's shop and transported to site for erection. The roof for Westminster Hall was made at Farnham in Surrey in 1395 and the cost of transport by wain to the Thames, where the timbers were transferred to lighters, was £19·07. The cost of carving angels on the hammerbeams varied from £0·75 to £1·33 each. An important feature of many secular roofs was the louvre, usually placed in the centre over the fireplace, where it was intended to act as a smokevent. Sometimes these louvres or vents were made by the tilers, to the indignation of the carpenters. Sometimes a barrel with the ends knocked out was used.

The main purpose of all roof structures is to support the weatherproof covering. In a previous section this has been dealt with in some detail. The cheapest form of roof was thack or thatch, often bought standing; at Bath an acre·of thatch was bought in 1420 for 8p, and in 1300 2·5 ha (6 acres) of rushes at Willingdon were bought for 90p for Pevensey Castle. The work of drawing straw or reed for the thatchers was usually per-formed by women, paid 1p a day for the work, the thatcher

Fig. 111. Placing joists

receiving 3p for his labour. After 1212, the use of clay tiles increased, the average price delivered to site being about 15p a thousand. Woolwich, a great centre for tilemaking, was supplying plain tiles in 1375 at 29p per thousand. Ridge tiles then, as now, were sold by the hundred at from 13p to 20p. Cornish and West Country slates were much cheaper at the quarry,

2p–5p a thousand, the carriage however greatly increased this price. Oak pegs for fixing tiles cost $\frac{1}{2}$p a thousand and in 1307 the rate for boring holes in stone slates for the pegs was 8p for 3000 stones. Moss for bedding the tiles was always paid for at the rate of $\frac{1}{2}$p a bundle, an indeterminate measure.

The mediaeval builder did not concern himself with the use of seasoned timber. Although green timber was in general use

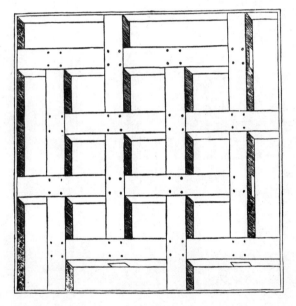

Fig. 112.   Timber frame

the slow pace of building must have allowed much of it to season naturally. It was generally appreciated that where it could be obtained seasoned timber was preferable. Oak was the usual timber preferred, being plentiful in most districts and easily transported by water. Being so much in demand good timber became scarce and by 1350 difficulty was being experienced in finding long lengths for beams. Oak was used not only for structural members but also for boards and laths. In

1343, 2p was paid for throwing an oak and 29p for cutting and sawing the tree into fifteen planks, each 7·30 m (24 ft) long. The production of laths was a minor industry, these being used to support roofing materials such as thatch and tile and as a ground for plastering. These laths were produced by splitting the wood, not by sawing as today. Their usual length was about 1·50 m (5 ft) and oak laths cost about 8p a thousand.

In addition to oak, a large quantity of timber was imported, mostly as boards from Baltic countries. These boards were

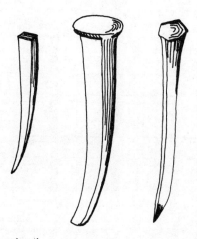

Fig. 113.   Mediaeval nails

often known as 'waynscottes' and being better seasoned than local timber were used for doors, screens and shutters. The timber was usually sold by the 'long hundred' measurement of 120, the price at Hampton Court in 1533 being 5p each. While much of this was oak, fir and deal boards were also imported and used for general work, but in the main fir was used for scaffolding and ladders. Native beech was used in large quantities, both as laths, in fittings and in the manufacture of centering for arches and vaults. Elm was in constant demand for piles, but when large oak became scarce, elm boards were used for less important work.

Apart from structural work, mediaeval carpenters were called upon to construct the fittings and finishings which go to complete a building. Whereas stone buildings had stone door frames, timber buildings had wooden ones, and the doors, as we have seen, were often made from imported oak. These were always properly framed and braced or ledged. A large door was often provided with a small 'wicket' or pass door incorporated in it, a 15th cent. example being the West door of Westminster Abbey. Window shutters were constructed in a

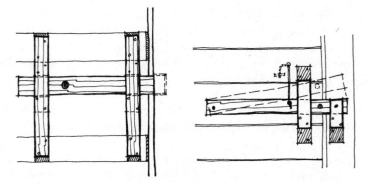

Fig. 114. Mediaeval timber door latches

similar way to doors, being called 'fenestra'; in 1364, carpenters were paid £16·96 for 407 doors and shutters for Sheppey Castle. Panelling was common in rooms from early times, always carried out in oak, usually imported for the purpose and often gilded and painted. Much of the early work was plain, but more elaborate designs later came into fashion, the gallery of Westminster Palace being provided with 914 m (827 yd) of 'draperie pannell' (linenfold) in 1532 at 9p for 914 mm (1 yd). Shields were carved on the panelling of the hall of Queen's College Cambridge, for 8p each and cresting was provided by the carvers at 3p a yard run. The main feature of the mediaeval hall, the screen, was also made by carpenters, and 'paper walls' or matchboard partitions were sometimes provided to divide up the larger rooms.

Iron was used in considerable quantity in building, not only for hinges, fastenings and laybars for windows, but also for constructional purposes, particularly iron ties. Iron used was either native or imported, the latter mainly from Spain through Winchelsea in Sussex and Sandwich in Kent. Gloucestershire and Wealden iron was well known and although imported iron was more expensive, 50p for 50·79 kg (1 cwt) against 35p for the local product in 1370, its continued use suggests that it was of better quality. Steel was necessary for the cutting edges of tools. Its price was so high that the usual method was to weld a piece of steel to the edge of an iron tool, thus reducing the cost.

Iron fastenings and ironmongery were made by smiths in many styles and types. Locks and padlocks were common, and hinges and latches were made to order as required. Many of these survive today, supporting and embellishing doors of the period. The usual method for protecting the metal from corrosion was by varnishing, although ironwork was sometimes deliberately blacked by the application of pitch, or whitened by dipping it in tin.

Great quantities of nails were required and used, each size being favoured with a distinctive name from 'spyking' to 'brodd', the latter being a small headless nail known today as a brad. In 1327 brads cost ½p a hundred, while 'spyking' nails cost 1p. Nails were therefore comparatively expensive items in the building costs, all being made individually by hand, a craft which survived into the 19th cent.

PART FOUR

Growth, Urbanisation, Industrialisation and Legislation

CHAPTER EIGHTEEN

Urban Building

The germ of the first city was the ceremonial meeting place of Palaeolithic Man, a site to which families or clans were drawn at intervals because of its spiritual or mystical associations. Some of the features of the city were present in the form of simple structures or earthworks.

To early man the main preoccupations were hunting and food-gathering. The first permanent settlements date from about 13,000 BC, and it was at this time that the first clearings for agricultural purposes were made. This process of domestication entered a second phase some 3000 years later when the systematic gathering of cultured seeds for food and the use of herd animals for power and transport, as well as food, became common. The village with its garden plots and fields formed a new kind of settlement, a permanent association of people and animals, of houses, barns and storage pits. This act of settlement helped to make agriculture self-sustaining, by utilising animal refuse for the rejuvenation of the soil.

The transition from the Neolithic village to the fully developed city took centuries, if not millenia. The oldest of these cities, excepting perhaps Jericho, date from around 3000 BC, the period in which evolved the potter's wheel, the plough, the loom, copper-smelting, the calendar, writing and the production of grain. The technological expansion of this period is only paralleled by that of our own era. The many diverse elements of a once scattered community were collected behind massive defensive walls. Vast labour resources were used in the construction of canals, temples, palaces and pyramids. In addition to providing the greatest possible protection to its inhabitants, the city offered the greatest opportunity to aggressive behaviour. While offering unparalleled freedom it also imposed a regulation of life which has, over the years, become second nature to man. Law and order supplanted brute force as the city took form around the Royal Citadel.

It is hard to estimate the population of ancient cities, although their extent can usually be verified by excavation. Khorsabad in Assyria, about 700 BC, enclosed some 300 ha (740 acres); Nineveh, a century later, 728 ha (1800 acres). Babylon, before its destruction by the Persians, was surrounded by a wall at least 17·70 km (11 miles) in circumference. It has been estimated that the average population of the ancient city was about 5000, which gave it the atmosphere and social organisation of a neighbourhood unit. The number of people to a room is impossible to verify but a modest Greek city house of about 200 BC measured 8 × 6 m (26 × 20 ft), about the same size as a 19th cent. house in an English industrial suburb. This compares favourably with a five-person house to Parker Morris standards today. The consistency of these figures over millenia is surprising.

Early cities were either restricted to pedestrian limits or kept within hearing distance of the main system of communication. In the Middle Ages the City of London was restricted to the extent of the sound of Bow Bell and this was an effective limit to urban growth at that time. The gossip of the town pump, the talk of the tavern, the proclamation of messenger and crier were the units of mass communication. The size of the city was restricted by the range of these communication media.

City development in the Western world began in Crete. Here the fertile soil supported a high level of agricultural activities and, times being peaceful, fortifications were non-existent. The city of Knossos, the hub of Cretan civilization, was unprotected and from this we deduce that the rulers were possessors of sea power, capable of protecting not only the island, but also the mercantile argosies which sailed to and from the ports. Between the 8th and 6th cents BC a new urban concept began to develop in the Eastern Mediterranean. A great number of small cities were founded, nourished by Greek cultural ideas. These cities were generally small and tended towards colonialisation rather than peripheral growth. This system produced great commercial centres, such as Rhodes, whose limiting size was dependent on the local countryside for food and building materials. Certain cities grew to dominate others, notably Athens.

The daily activities of the Greek city took place out of doors. The centre was formed by the Acropolis: not the castle or palace of former civilizations, but the temple, constructed in accordance with tradition and housing the sculptural image of a god or goddess. Even in the larger cities the temple was not of sufficient size to house the whole population, consequently the principal ceremonies were conducted outside the building, within the sacred enclosure. In its temples and monuments, magnificent as they were, the Greek city was not unique. Its real importance lay in its adjustment to human scale and proportion, being neither too large nor too small for the fundamental urban requirements of co-operation and communion.

According to the tradition, Rome developed by the union of various tribes from the neighbouring hills, and the symbol of this union was the foundation of a common meeting-place or forum. This was not simply an open square. As it developed, the forum was more of a precinct in which shrines and temples, halls of justice and council houses all played a part. The open spaces allowed orators to address large crowds, and in bad weather large basilicas or halls served many purposes. The Roman forum was the centre of public life, but elsewhere the city was a jumble of narrow passages lined with shops and taverns and overshadowed by high tenements. The Roman contribution to urban hygiene was the bath. Free public baths were provided in 33 BC: vast enclosures, one monumental hall leading to the next, hot, tepid and cold baths, rooms for massage and refreshment, gymnasia and libraries. These buildings are among the great architectural achievements of Rome, and wherever the Roman went he took the idea of the public bath with him. The habit of bathing must have tended to reduce the hygienic and sanitary hazards of the time.

The principal problem of the Roman was, however, to employ the leisure granted to him by an indulgent hierarchy. The need for mass entertainment became imperative as Roman city life became more artificial. The circus, a banked enclosure open to the sky where Romans in their thousands gathered to

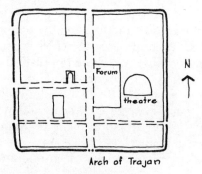

Fig. 115. Typical Roman town plan: Timgad

view the spectacles, the amphitheatres where mass cruelty and murder were staged for the entertainment of the masses, were urban developments which sprang from this need.

Rome was the antithesis of the Greek city. After its enclosure by the Aurelian Wall in AD 274 the area contained covered 1223 ha (3323 acres) with, in addition, some 667 ha (1600 acres) of built up area outside the wall: a formidable urban complex. What was lacking in Rome was a system of control which, incorporated with her talent for law and system, might have led to her survival. Rome's main contribution to city development must therefore be regarded as that of 'over growth', the loss of human scale – a lesson found hard to assimilate by later conurbations which have regarded physical and economic expansion as synonymous with prosperity and culture.

The early mediaeval town was still part of the open country. Most of the population had private gardens and at harvest and holiday times migrated to the country. Although most houses were built in rows for cheapness as well as warmth, it was usual to place the long elevation to the street, increasing the width of the garden plot at the rear. Town plots were generally about 30 m (100 ft) deep with a frontage of 15 m (50 ft).

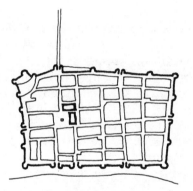

Fig. 116.   Mediaeval fortified town: Aigues-Mortes. *c.* 1300

Sometimes the rectangular block was coupled with a corresponding outline to the city as a whole, enclosed within an encircling boundary wall. In some instances, mediaeval plans were more irregular than modern, probably due to the exploitation of uneven sites for their defensive advantages. Mediaeval builders had no love of symmetry for its own sake; it was simpler to build to the contours than to grade the site to suit the development. Many of the irregularities in the plans are due to streams now covered over, trees which have been felled or the boundaries of old field systems.

The determining elements in the mediaeval town were the enclosing wall and the central open space. Here was located the principal church, the town hall, the market and the inns. The main thoroughfares were planned to facilitate quick and easy movement to and from the main gates, and converged on the church at the centre. The planning of the market place beside the church was deliberate. It was in the church that the citizens most frequently assembled and it was here that arms were stored. In time of trouble, valuables were deposited in the church for safekeeping. The church was a community centre and at times of great public festivals could also serve as a refectory.

Except for the church, the mediaeval planner kept to human scale. Almshouses were for small communities of seven to ten persons, hospitals were for districts of from two to three thousand people. The decentralisation of social services not only reduced needless movement but kept the whole town in scale. The early mediaeval street was a line of communication rather than a means of transportation. These unpaved streets were sometimes narrow and the overhang of adjoining houses afforded protection to the pedestrian (see the Shambles at York). Sometimes buildings were constructed to form a covered walk or arcade such as have survived at Chester and Ledbury. This provided protection not only to the pedestrian but also to the stalls and booths of merchants and shop-keepers, usually open to the weather.

Between the 15th and 18th cents the form and content of urban life was transformed. The new pattern of life sprang from a changed economy, that of mercantile capitalism, and a new political framework, the national state. The disintegration of the mediaeval synthesis was due, in the main, to the Black Death in the 14th cent. A contributory factor was the decline in the force of the moral and social consensus which had been a feature of the late Middle Ages. Two powers were fighting for supremacy, the Crown and the Municipality. The 14th cent. state required a permanent bureaucracy and courts of justice, centrally located to conduct official business. The centralisation of authority created a capital city which, commanding the routes of trade and military movement, was in its turn an important contributory factor in the unification of the state.

All this caused a loss of power in local centres; economic privileges, once conferred by the city, became the perquisite of royal power. The population of the capital city increased rapidly because of the privileges it could confer on its occupants, and the founding of new towns and cities ceased. City building was no longer a means of achieving freedom and security, it was a means of achieving political power.

Up to the 15th cent. defence was stronger than attack. The improvements in the production of cannon and the invention of the iron ball changed this. While the defender's cannon

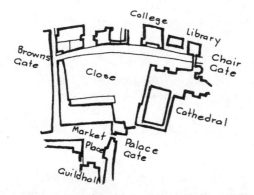

Fig. 117.   Mediaeval cathedral square: Wells, Somerset

could do comparatively little harm to companies of men in open formation, the effect of the attacker's artillery on the stone walls and tiled roofs of the city was disastrous. By the end of this century mediaeval walled cities were completely vulnerable to the new forms of attack.

New stystems of fortification were devised with outworks, salients and bastions in star formation, which for two centuries protected the cities of Europe. The cost of providing and maintaining these works cast a financial burden on the inhabitants which may well have contributed to the decline in living standards which ensued during this epoch. The restricting effect of the new fortifications was a planning disaster. New growth could only take place either by utilising the gardens and open spaces which were formerly so valuable as a safeguard to the health of the population, or by increasing the height of buildings (up to 10 or 12 storeys in places such as Edinburgh). The intensive development of the art of fortification altered the pedestrian scene of the mediaeval city to the expanding world of the Baroque city with its long range artillery and wheeled vehicles. The army became a new factor in city life, the barracks took the place of the mediaeval monastery, and the parade ground was conspicuous in the new city plan.

The avenue was the most important feature of the Baroque city. With technical improvements which replaced the solid

Fig. 118.   Plan of Baroque city from *Architecture of Fortifications* by Daniel Speckle. 1589

wheel with one of separate parts – hub, rim and spokes – came a more general use of wheeled vehicles. The introduction of vehicles was resisted; mediaeval streets were not suitable for their use. Movement in a straight line was economy, and where it was not possible to design a whole new city to the Baroque plan, the layout of a few new avenues could redefine its character. In addition, avenues were a great military asset, not only for the rapid movement of troops from one section of the city to another, but also contributing to the sense of power imparted by the marching column. The buildings on either side formed a setting, a place where spectators gathered to watch the parade. Whereas in mediaeval days rich and poor jostled together on the street, now the rich rolled along the axis of the avenue and the poor kept to the gutter or that new innovation, the pavement.

The military parade had its feminine counterpart, the shopping parade. The old open market was generally restricted to provisions, except in the poorer quarters. The open shop with a workroom at the rear began to disappear, being replaced by a new type of retail dispensary, carefully protected from the elements by glass windows. These were enlarged to extend over the whole of the ground floor frontage and served as a centre for display, the interiors being smartly decorated and fitted out for the enticement of custom. While the special market day lingered on in rural communities, in the commercial town every day was a shopping day. The producer became as anonymous as the consumer, the shopkeeper or middleman coming to the fore by anticipating the requirements or by manipulating the taste or judgement of the public. In the great cities, too large for people to know their neighbours, the standards of the market and the shopkeeper generally prevailed.

The influence of the court bore upon the city in every aspect of its life. Under its patronage the theatre took its modern form, when spectators became seated as passive observers of the drama seen through the proscenium opening. The great collections of art which form the basis of the Musée du Louvre

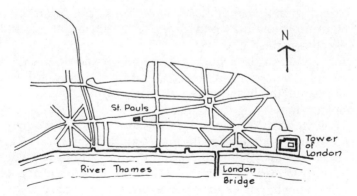

Fig. 119.   Sir Christopher Wren's plan for rebuilding the City of London. 1666

G

in Paris and the National Gallery in London were started; and at a time when music was dying in the home due to rising distinctions between master and servant, the orchestra and its natural home, the concert hall, first made their appearance. All these institutions made their mark on the city plan, sometimes under private patronage, sometimes with royal or municipal support. The city park made its appearance, the equivalent of the pleasure ground and playing field of the mediaeval city. The idea of the open space was not wholly devoted to the good of the general population. As in Regent's Park, London, the values of the ground rents of the adjacent Crown Lands greatly increased in value by its proximity. Pleasure gardens such as Ranelagh and Vauxhall in London, influenced by court ideas and manners, were provided for the population at prices they could afford, the Beer Garden being the German equivalent. These institutions gradually degenerated into the great international fairs and amusement parks which are a feature of certain cities today.

On the cultural side, in addition to the theatre and the concert hall, museums, proceeding from natural curiosity or acquisitiveness, began with collections of coins, extending later to natural history. The opening of the British Museum in 1759 was a landmark, when the museum ceased to be an amateur institution and began to advance along the path of intellectual and scientific investigation, and public entertainment. The lions of the Tower of London were a curiosity for several centuries from mediaeval times. The extension of similar collections, their housing and exhibition, took place at the same period as the founding of museums and flourished not only as a depository of the trophies of explorers and hunters but also as the headquarters of many earnest and enquiring scientific minds.

For the mass of the population, however, the Baroque city was a deterioration of environment and a depression of their standard of life. The great centres began to absorb population with no attempt at limitation. While the cities of Europe were either decaying or being transformed into a mechanical pattern, the countryside was undergoing rebirth and improvement. The fine stone villages of the limestone belt of the

Cotswolds and the elegant clapboarded houses of the Weald of South East England date from this period. The unification of the three field system of manorial days into larger land parcels integrated the landscape. The wild forest became the park, hand industries escaping from the commercialisation of the cities found refuge in the villages. The way of life and organisation of the mediaeval city took root in the new villages which today in many cases preserve the layout of the older city unit.

During the 18th and 19th cents, the people of Western Europe opened up vast areas of the earth's surface where land-hungry masses could settle. This migration brought to European farming new crops – maize and potatoes. A surplus of grain made possible the bulk distillation of whiskey and gin. These supplemented the inadequacy of heating arrangements with internal warmth. The increase in food supplies made possible an increase in population, villages expanded into towns, towns into cities. Men built in haste, newcomers crowded into whatever habitation they could obtain.

The two main elements in the new urban complex were the factory and the slum. The factory was the nucleus, every other detail of life was subservient to it. Utilities, water supply, sewage disposal, all were afterthoughts. The factory was provided with the best site, often near a waterfront. Here water could be obtained for the boilers and the waste products of the process discharged. The transformation of watercourses into open sewers has not been resolved today. Great mounds of solid waste, slag and rubbish, blocked the view. No authority existed to zone or concentrate factories, or to plan housing in appropriate areas. The juxtaposition of factories, commercial premises and dwellings was a feature of all industrial cities. Only physical features could restrict the spread of industrial plant, the valley bottoms being more convenient in most areas. In the old cities, housing deteriorated into the one room family unit, provided in converted one-family houses of former days. A degree of dirt and squalor was reached which could never have been equalled in mediaeval Europe. In England, back to back housing provided living units by the thousand, many rooms lacking free ventilation or natural daylight, toilets few and far between. In Manchester, one district in 1844 boasted only one toilet to every 212 people. In Liverpool, one person in six lived in an underground cellar. In the new industrial towns, municipal service was almost wholly absent, whole districts being without even well water.

The depression of city environment was not restricted to the dwellings of the factory employees. The middle class housing of the 19th cent. provided quarters for the staff which were of a similar low order and many of the principal rooms of the owners faced onto deep airless courts. Typical urban housing of the century was of two types, tall blocks in cities such as we have seen in Edinburgh, Paris and Berlin, and two storey buildings as in London and Chicago. All are linked with common characteristics, block after block repeats the same pattern, the same formula of drabness and uniformity. Lack of orientation and open spaces are endemic. The age of invention and mass production by-passed the housing needs of the population. The standardisation of the factory slum was the chief urban achievement of this period. The organisation of the layout into rectangular blocks presented few problems. With tee and set square the municipal engineer, without any training as architect or sociologist, could plan a metropolis. The rectangular plan is often considered to be America's legacy to the world. In fact it originated in the 17th cent. in the growing cities of central Europe, such as Berlin. This layout was perfect for the speculator. Plans could be made for unlimited future extension without thought to future consequences. No consideration need be given to prevailing winds, foundations, aspect or any of the other factors necessary for the proper utilisation of a site

For the middle classes, the escape from urban disaster took the form of building villas in the suburbs. These already existed in the 16th cent. and in times of plague the well-to-do citizens of London always dispatched their families to the clean air of villages such as Hampstead and Clapham. At first, the possibilities of suburban living was restricted to those with their own transportation but by 1820 in London the new villa districts of Barnes, St John's Wood and later Bedford Park were taking shape. These suburbs followed romantic principles, streets followed contrived curves, and contours were incorporated for the sake of irregularity. Simple natural forms of trees, gardens and open spaces were cheaper than the extravagances of wide roads and paving. The popularity of the movement resulted in the rise of land values, the city crept out towards the suburb and its qualities decayed. The only real benefactors of suburban development were a few industrialists, Salt at Saltaire, Cadbury at Bourneville and Lever at Port

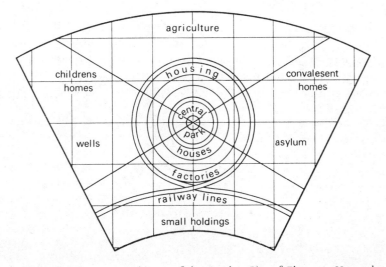

Fig. 120.   Diagrammatic layout of the Garden City of Ebenezer Howard. 1898

Sunlight, all of whom made significant contributions to urban life.

The earliest contributor to the housing movement was Robert Owen. Owen proposed to build small balanced communities in open country to assist factory workers to rise above the conditions in which they lived under the factory system. His plan was superior to any other urban pattern in the 19th cent. as it provided, as part of the concept, some of the essentials of social life. Owen's work compares very favourably with work done in the 20th cent., especially as his projects were not marred by the quaint and the picturesque. These new communities were directly related to industry and they formed small civic units. Their weakness was that they owed their existence to the initiative of the enlightened employer.

By the end of the 19th cent., the conception of the garden city, a new centre deliberately contrived to provide a balanced environment, gained impetus. In 1904 the first fully equipped garden city, Letchworth, was founded by a specially organised public utility association. Its growth was slow, but steady, and with its success in incorporating industries and workers' houses, it became a focal point in the minds of planners and administrators throughout the world. The fulfilment of its founders' dream can be seen in the new towns which have been created throughout the British Isles since 1945 and for which this country has received much merited praise.

CHAPTER NINETEEN

The Industrial Revolution

The Industrial Revolution had an important effect on the building industry in Britain. In the period up to the beginning of the 17th cent. the economy had been largely agricultural and the staple industries had been dependent on wool. All associated processes such as spinning and weaving were 'cottage' industries. These were concentrated largely in localised areas of the West Country and the North of England. The soil in certain areas of the West Riding of Yorkshire provided only a partial livelihood for farmers, arable land was scarce and the acid soil and harsh climate unsuitable for cereals. There was, however, ample room for large flocks of sheep, and it was inevitable that, with heavy rainfall on the surrounding hills providing ample supplies of soft water to wash and scour the wool, the natural thrift of the local inhabitants would result in local processing of the wool. Many farmhouses in this area originally housed these cottage industries, where the whole process from clipping the fleece, through washing, picking, carding, spinning and weaving was carried out. These farmhouses are usually distinguished by a long row of mullioned windows on the upper floor which provided ample light for the weavers. Wool and cloth were transported by pack horse over the moorland paths to be sold at the Cloth Halls of Leeds,

Huddersfield, Bradford and Halifax. The only survivor is at Halifax, where the building known as the Piece Hall was erected in 1779. This building contains 315 rooms, each originally occupied by a separate manufacturer or merchant. It is an imposing stone structure, three stories high and built around a central courtyard covering an area of over 8361 m² (90,000 ft²).

As we have already seen, the building industry had for centuries concentrated on the construction of castles, royal palaces, church buildings, schools and colleges, houses and mansions for the growing numbers of wealthy patrons, market halls, hostelries and public buildings. The advent of industrialisation brought about many changes in this pattern of work.

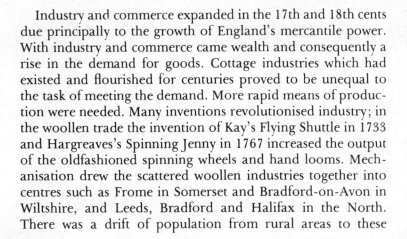

Fig. 121.   The Piece Hall, Halifax. 1779

Industry and commerce expanded in the 17th and 18th cents due principally to the growth of England's mercantile power. With industry and commerce came wealth and consequently a rise in the demand for goods. Cottage industries which had existed and flourished for centuries proved to be unequal to the task of meeting the demand. More rapid means of production were needed. Many inventions revolutionised industry; in the woollen trade the invention of Kay's Flying Shuttle in 1733 and Hargreaves's Spinning Jenny in 1767 increased the output of the oldfashioned spinning wheels and hand looms. Mechanisation drew the scattered woollen industries together into centres such as Frome in Somerset and Bradford-on-Avon in Wiltshire, and Leeds, Bradford and Halifax in the North. There was a drift of population from rural areas to these centres of employment, and the building industry was kept busy erecting new mills beside the rivers and streams so necessary for processing of wool, and in building cottages for the workpeople. The silk industry had followed a similar expansion, but this was primarily due to external political and religious pressures. After the Revocation of the Edict of Nantes in 1685 by Louis XIV, more than 100,000 Huguenot refugees fled from France. Many settled in England in provincial towns such as Canterbury, and also in London where they established the silk industry at Spitalfields. In 1718, the first silk mill to be driven by water power was erected in Derby. So long as power was derived from water wheels, mills were located towards the headwaters of streams and rivers where sufficient flow of water could be more easily compounded. As there was little smoke or grime, the houses of the workers were built close to the mills.

The invention of the steam engine altered all this. Matthew Boulton and James Watt produced the first steam engine in their Birmingham factory in 1775 and these engines were used in collieries, operating the winding gear. In 1785 they were first installed in a cotton mill and from this time onward mills and factories were erected in any locality where fuel was readily available. Large areas in the Midlands and the North close to coalfields, previously predominantly agricultural, became intensely industrial with a corresponding rise in population.

The widespread use of the steam engine necessitated radical alterations in the system of transport and distribution of coal. Although the design and construction of roads under the guidance of men like Macadam and Blind Jack Metcalfe of Knaresborough improved, journeys were still very slow. Pack-horses were used for carrying coal for short distances and wagons were used for the distribution of merchandise. Attempts to force local inhabitants to repair roads within their parish were ineffectual. The first Turnpike Act of 1663 introduced the principle of levying tolls on road users for the repair and improvement of roads but as late as 1753 it still took ten days to travel from London to Liverpool.

It was therefore by navigable rivers, and later by canal, that the cheapest and easiest method was found for the transportation of coal and bulky building materials to the new industrial areas. At the beginning of the 17th cent. there were about 700 miles of navigable rivers, mostly in Southern England. The main carriage was the transportation of 'sea-coal', with return cargoes of timber, stone and grain. The first application of the pound lock to improve river navigation was at Exeter, when the Exeter Canal was opened in 1567 to allow barges to pass around a weir' to reach the city. Another was constructed at Waltham Abbey. Although a few more locks were built in the 17th cent., it was not until the 18th that the use of locks on rivers became common.

The first Canal Act was passed in 1755 permitting the construction of the Sankey Canal linking the Wigan coalfield to the River Weaver, allowing coal to be transported direct to Liverpool and the salt works of Cheshire. It was, however, the third Duke of Bridgewater who, by constructing a canal to link a colliery on his estate at Worsley to Manchester, initiated the

Fig. 122. Pont-y-Cysylltau aqueduct carrying the Llangollen canal over the River Dee. Thomas Telford, 1805

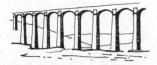

rapid spread and use of canals. By the end of the first quarter of the 19th cent., there were over 3000 miles of canals operating in England.

In the 18th and 19th cents a vast amount of construction work was carried out. The population doubled in 100 years to 12 million in 1800, increasing to 22 million by 1850. Construction works included new roads, bridges and aqueducts, canals and docks, wharves, factories, mills and warehouses, vast housing estates, innumerable civic and secular buildings, churches and chapels. The large scale construction work which we would today call civil engineering was something new to the industry. It brought into being the civil engineer who specialised in the problems raised by these works and advised in a consultative capacity on their practical solution. Some of these men became famous, men such as James Brindley and James Watt for their work in connection with canals, Thomas Telford for his roads, bridges and canals, John Rennie for his bridges and aqueducts and John Smeaton for his marine structures and lighthouses. In 1779 the first wholly iron bridge was thrown over the River Severn at Coalbrookdale. With a clear span of 30 m (100 ft) it is still in use today.

Great numbers of men were needed for these works. Those employed on the new inland waterways were known as navigators or 'navvies', a term with us today. The amount of labour involved is impossible to describe, and it was nearly all carried out using pick and shovel. The aqueduct carrying the Llangollen Canal over the River Dee is considered to be Thomas Telford's masterpiece. Built of stone, 38 m (126 ft) high and 305 m (1000 ft) long, the canal is carried in a trough of iron 3·65 m (12 ft) wide with a single tow path. Completed in 1805, it is still in use today. Many canals passed through tunnels in

Fig. 123.   Iron bridge over the River Severn, Coalbrookdale. 1777–9

hilly country, and the longest, the Standedge Tunnel between Yorkshire and Lancashire, is over three miles long. The roof is so low that barges were 'walked through', by the bargee lying on his back on the deck and pushing against the roof with his feet.

A vast amount of building work was in progress in the new industrial centres. The invention of the steam engine and the easy transportation of coal enabled the establishment of factories and mills on any suitable site. Factories sprawled over the areas, hemmed in by workers' houses in indescribable congestion. Drainage, often poor or non-existent, consequently produced appalling sanitary conditions. Advantage was taken by speculators of the pressing need for housing. With very little knowledge of building standards, vast fortunes were amassed by the erection of these slum properties. The social and economic conditions of the workpeople became increasingly difficult and the seeds of social unrest in the next century were sown.

Although steam engines were produced as far back as 1775, it was not until 1830 that the first railway with steam loco-motives was opened. Previously, there had been many tram-ways connecting quarries and pits with factories and wharves, the motive power being provided either by horses or gravity. In the next ten years, London was connected to Liverpool and Bristol by rail, the engineers being Stephenson and Brunel respectively. Railways spread across the country with increas-

Fig. 124.   The Britannia Bridge, Menai Strait. Robert Stephenson, 1849

Fig. 125.   Clifton Suspension Bridge. I. K. Brunel, 1860

ing speed and brought great changes in building techniques. The effect on the use of local indigenous materials was disas-trous. Welsh slates, hard red Midland bricks and later cor-rugated iron and corrugated cement sheets flooded in to every corner of the country, and one by one the local materials began to disappear. All regional styles were destroyed except where the well-to-do could either afford or were willing to continue the old traditions with the aid of an architect. What had been the local idiom of the people became the plaything of the wealthy.

The latter part of the 18th and early years of the 19th cents were years of great opportunity for the building industry.

Mention has already been made of the unprincipled exploitation of sections of the community by mercenary interests. Wealth was coming to the merchant and professional classes who needed new homes within reasonable distance of their business. Landowners contracted to develop new housing and some builders and architects raised capital to develop on their own account. Excellently designed developments were carried out in London, the spa towns of Bath and Cheltenham, and the seaside towns of Weymouth and Brighton.

In London, the most successful speculative builder was Thomas Cubitt, (1788–1856) who after starting work as a journeyman carpenter started business on his own with a workshop in Gray's Inn Road. In 1824 he contracted with the Duke of Bedford to develop the area around Tavistock Square, and afterwards with the Duke of Westminster and Mr Lowndes for the estate which we now know as Belgravia. In this, Belgrave Square is considered the finest square in Britain today. Between 1830 and 1840, Cubitt built many hundreds of stucco fronted houses along spacious streets and squares in the area south of Belgravia. Sometimes he worked in collaboration with an architect, often the design was carried out by his brother, Lewis. Cubitt also built Osborne House on the Isle of Wight in collaboration with the Prince Consort.

The work of the Woods, father and son, architects in Bath during 18th cent., is outstanding. Financed by local wealthy landowners, they began to build in 1728 and their work shows great skill and elegance. Particularly noteworthy is the Circus, completed by Wood the Elder in 1754 and incorporating houses embellished with the three classical orders, and the Royal Crescent built by Wood the Younger in 1769 of houses adorned by a colonnade of Ionic columns.

Cheltenham, another inland spa, became fashionable after the visit of George III in 1788 and has many fine buildings. Several of these are situated in the district known as Pittville, built on a magnificent scale by Joseph Pitt in the early years of the 19th cent. At Brighton, the interest shown by the Prince Regent and his conversion of a small house into what we now know as the Royal Pavilion, under the supervision of John Nash, led to considerable development in the town. This has given us such fine works as the Royal Crescent and Brunswick Square in Hove. A speculative builder, Thomas Kemp, erected a large estate to the East of the town which is known today as Kemptown. John Carr of York and James Essex of Cambridge were also responsible for much work of quality during this period.

It seems curious to us today that in an era when progress in industrialisation moved with an impetus never before reached, industrial techniques were totally disregarded in the field of the building industry. Probably the greatest exception was the vast 'greenhouse' erected by Joseph Paxton in Hyde Park in London in 1851 for the Great Exhibition. One can only assume that not only social and economic factors but also the continuing tradition of certain craft work enabled craft construction to continue despite the industrialisation of every other facet of life.

CHAPTER TWENTY

Drainage and Public Health

The removal and disposal of human and domestic waste has been a problem ever since man began to settle in communities and adopt his present mode of existence. Neolithic Man excavated pits into which refuse and rubbish was thrown and many of these have been discovered to provide interesting information about his domestic habits. Streams and watercourses were also used to convey waste material away from the occupation site, and where these were at some distance, ditches and drainage channels were constructed to serve as open sewers. Remains of these have been discovered at Khorsabad (800 BC) constructed of brick and arched over with a brick vault. Remains of a foul drainage system have been excavated at Knossos in Crete, and it must be assumed that the gutter which runs down the centre of most paved Greek and Roman streets served a similar purpose when regularly flushed by stormwater. Many of the larger Greek and Roman houses were provided with their own drainage systems which were, in the principal centres, connected to municipal sewers conveying both foul and storm water to discharge either into natural watercourses or, in the case of a coastal town, into the harbour.

In Italy, the peasants of Southern Etruria found it necessary, from early times, to excavate drainage channels in the local volcanic soil to clear the swamps which formed so easily in this region. The first settlements forming the future city of Rome were constructed on the five hills, and expansion into the valleys was not possible until the streams which ran through them had been canalised. This work was carried out, only roughly at first, the great drain of Rome dating from about 578 BC and known as the Cloaca Maxima remaining open to the sky for several centuries. Its first covering was of wood planks, and when it was vaulted towards the end of the Republic, the work was carried out in a massive manner, the 3·35 m (11 ft) diameter span being arched over with three rings of voussoirs, each 760 mm (30 in) high.

In mediaeval times the proper disposal of household refuse and filth received little attention. As we find later, in connection with water services, many of the monastic foundations were better equipped in this respect than lay buildings. At Christ Church, Canterbury, the waste from the water supply and all stormwater was carried up under the 'necessarium' or latrine to flush the drain. This was so common a feature of monastic planning that it often determined the placing and layout of the conventual buildings on site. At St Albans a stone cistern was provided to store rainwater used to flush the latrine, and at Queenhythe the 'necessary house' provided for the use of the citizens was extended to give access to the Thames. It was very common in those days to provide these structures over or immediately adjacent to a natural watercourse. In moated castles or manor houses the usual provision was either a simple chute leading out at the foot of the wall or a corbelled projection, both discharging into the moat (The Gatehouse, Mickleham Priory, Sussex).

In towns and cities the normal provision for the disposal of sewage was a cesspit, these being provided in most mediaeval houses. The pits were often constructed of stone and most had a vaulted roof. They undoubtedly became offensive in use, especially in hot weather. It is interesting to note that on occasions ventilating shafts were provided to these cesspits, similar to modern practice. In the main, however, the accumulation of filth and waste matter caused by lack of proper sanitation and congestion led to repeated outbreaks of serious epidemics. The Black Death, an outbreak of bubonic plague which spread from the East and was contracted from infected rats, is estimated to have caused the death of one-third of the population of England in the years 1348 and 1349. Plague recurred on several occasions over the next three hundred years, as the population continued to live in insanitary and over-crowded conditions and the contamination of water supplies by overflowing cesspits aggravated conditions. Streams such as the River Fleet in London, polluted by filth and excrement, were little more than rat-infested sewers. By the end of the 16th cent. conditions were so bad and overcrowding of houses and tene-ments had created such pockets of congestion that the Crown had to take action. An Act of 1592 sought to restrict the building of further dwellings within the City of London and its environs and in 1602 a further Act was passed which forbade the sub-division of houses into tenements. James I issued other pro-clamations designed to restrict the growth of the city and the Act of 1607 required the issue of a licence before a house could be built within the city. The restrictions were maintained and extended over the next few years, the only effect being to aggravate conditions which the regulations were designed to cure. Charles I continued the policy of restriction instigated by Elizabeth and in addition called on Justices of the Peace and others with authority to imprison workmen engaged on new buildings within three miles of the gates of London. The policy was continued during the Commonwealth but the limit was increased to ten miles, this being reduced in 1661 by Charles II to the more practical limit of two miles.

The sanitary conditions in London continued to be un-satisfactory during the 18th cent. and the mortality rate con-tinued to be high. In the year 1764–1765 there were 23,000 deaths in London of which 35 per cent were infants under two years of age. In that year the Commissioners of Sewers and Pavements presented their report to the Lord Mayor, Alder-man and Common Council of the City of London and the following extracts make interesting reading.

1. The pavements and streets of London are roughly paved due to the custom for householders to effect their own repairs to their frontages.

2. The centre channel provided is dangerous and disagreeable to foot passengers and carriages alike.
3. The common practice of depositing household refuse in the centre channel of streets is dangerous to health.
4. The footways, not being raised above the street level, are often covered with mud.
5. Footways are obstructed with shop windows, showboards and cellar doors.
6. The increase in projecting signs to business and commercial premises obstructs the free circulation of air and intercepts the light of lamps.
7. Foot passengers are annoyed in wet weather by the discharge of water spouts from roofs.
8. Lack of street names and house numbers causes great inconvenience to strangers.

It is surprising to read of these serious criticisms of the conditions of streets and footways in the greatest mercantile centre in the world.

During the early 19th cent. public conscience was stirred, and at last revolted against the insanitary conditions in towns and cities. Smallpox was the prevalent disease and in 1800 the death rate in London alone was one in twenty-four of the population. Sewage treatment, as we understand it today, did not exist and a drainage system was merely a pipe taking foul water by the shortest route to the nearest watercourse. It was not until the 1850s that proposals were made to intercept, by artificial channels and sewers, the daily flow of London's sewage and carry it to outfalls well downstream of the city. The Metropolitan Board of Works was created in 1855 to assume responsibility for London's sewers. A network of sewers was laid and the watercarriage system of disposal introduced, which led to a much improved standard of public health.

In the meantime, in 1848, the Public Health Act was passed and modern sanitary law came into being. The Town Improvement Clauses Act of 1847 had authorised the appointment 'of a person duly qualified to act as a local surveyor of paving, drainage, etc.' and this provision was embodied in the Public Health Act of 1875. So far building and sanitary legislation had been almost wholly confined to London, but it was appreciated that extension to other cities was long overdue. The Local Government Act of 1858 gave urban authorities powers to make bye-laws for the control of building, subject to Home Office approval. Subsequent Acts of Parliament have continued to extend and modernise legislation to deal with matters connected with public health, aspect of buildings, and urban and rural communities.

CHAPTER TWENTY-ONE

Building Legislation

The early towns, especially London, cramped within their encircling walls, were mostly composed of timber-framed houses, shops and warehouses huddled on either side of narrow streets. The supporting structures of these buildings were usually timber frames filled in with wattle and daub and the roofs were usually thatched with straw or reed. The danger from fire was very real and so many disastrous fires occurred that it became necessary for some authority to be created to take action. Roman London was destroyed by fire in AD 61 and again about AD 120. In AD 961 Saxon London was badly burnt. Other disastrous fires occurred in 1077, in 1087 when St Paul's was burnt, in 1136 when London Bridge was destroyed and again in the years 1161, 1212 and 1264. Other towns, Canterbury, Chichester and Worcester among them, were devastated between 1161 and 1202.

In 1189, the first Mayor of London, Henry Fitz-Alwyn, promulgated the London Assize, the first Code of bye-laws in England relating to town planning and the construction of buildings. To reduce the risk of fire spreading from one building to another, the code provided for the erection of party walls 900 mm (3 ft) in thickness and 4·9 m (16 ft) in height, with certain reductions in thickness to 300 mm (1 ft) for permitted alcoves. The distance from the ground to the under-side of the overhanging upper storeys was to be not less than 2·40 m (8 ft). Streets continued to be very narrow, in many cases the projecting upper storeys being but a few feet apart.

After a disastrous fire in 1212, an ordinance was issued by King John prohibiting ale-houses (being especially vulnerable to fire) in the City except by special licence, unless they were built of stone. Bakehouses and breweries were forbidden to use straw or reed as fuel. In addition their walls were to be plastered and whitewashed and all superfluous woodwork removed. In future no roofs were to be covered with any material but tile, shingles or lead and all existing thatched roofs were to be plastered on pain of demolition.

In spite of this legislation, London remained full of timber-framed buildings until 1666. Before the end of the 13th cent. when sea-coal was brought into London, the main fuel was charcoal and faggots. The open hearth was placed as near to the centre of the room as possible and the smoke and fumes allowed to escape as best they could through outlets in the roof, smoke vents or controllable louvres. There were few chimneys except in stone built houses.

After the Dissolution of the Monasteries, monastic land was quickly developed and a great deal of building took place outside the city walls, incorporating the surrounding villages into the metropolis. Thus in the reigns of Elizabeth I, James I, Charles I, and during the Commonwealth, the City of London was greatly enlarged. Several decrees were issued during this period which sought to restrict building develop-ment but they had little success and fire remained a serious threat.

The Great Fire of London in September 1666 destroyed over 13,000 houses. Within ten days Charles II promulgated an Act of Parliament setting up the Court of Fire Judges. This Act made provision for hearing disputes between those parties affected by the fire to consider how each should bear a proportion of the loss according to their several interests. The accompanying Act for the Rebuilding of the City of London required that all new buildings should be of brick or stone, and, in addition, were to have proper party walls to restrict the spread of fire. This Act of 1667 may be regarded as the first comprehensive building regulation for London.

This Act allowed four types of houses. The first 'least sort of building in bye lanes' was restricted to two storeys in height excluding cellars and attics, the second 'for streets and lanes of note' was restricted to three storeys, the third fronting 'high streets and principal streets' to four storeys and the last, being for merchants and mansion houses of large size, which were not built to the street frontage, were not to exceed four storeys in height. Many details of construction were laid down, the heights of rooms, the thickness of external brick and party walls, the depths of cellars and the sizes of supporting timbers. The position of timber in relation to chimneys and flues was carefully defined.

The task of rebuilding the City was a tremendous undertaking, calling for many thousands of craftsmen. The guild members of the City were too few in number to deal with the influx of work and in consequence 'foreign' workmen from other parts of the country flooded into London. Normally these men would not have been permitted to work in the capital but special dispensation was written into the Act of 1667. In this the guild privileges were put aside and it was not long before there was trouble. A petition from the guilds to the Court of Aldermen in 1670 complained bitterly of the violation of their ancient usages and customs; this appeal, however, was not very successful.

To raise funds for the acquisition of land to effect street widening and to help pay for the rebuilding of St Paul's, city churches and public buildings, the Act imposed a tax of 5p on every 'chaldron' of sea-coal landed in London for a period of ten years. By the Additional Building Act of 1670 the tax was increased to 10p and the period extended to 1687.

Additional legislation to attempt to reduce the risk of fires was enacted in subsequent years. In 1707, an Act prohibited the construction in London of overhanging eaves and cornices, and required that walls should be carried up as parapets to a height not less than 457 mm (18 in) above the roof. In 1708 it was required that chimney backs and jambs were to be at least 229 mm (9 in) thick with at least 115 mm ($4\frac{1}{2}$ in) of masonry or brickwork between flues. These latter were to be parged and fireplace openings arched in brickwork. No woodwork was to be placed within the structure less than 127 mm (5 in) from a flue.

In 1724 it was required that notice should be given to an adjoining owner of intention to rebuild a party wall. No openings were permitted in party walls unless the adjoining buildings were to be used as one and no timber was allowed to be built into any party wall. The Act of 1760 increased the thickness of party walls in cellars from two to two and one-half bricks and above this height from one and a half to two bricks in thickness, respectively. Timbers were permitted so long as they did not penetrate more than 305 mm (12 in) and were so staggered that the distance between adjacent beams was not less than 229 mm (9 in).

The Act of 1764 prohibited the use of timber under any hearth and increased the distance timber could be placed from a flue from 127 mm (5 in) to 229 mm (9 in). As can be seen from these regulations, an immense amount of energy and application was put to the problems of reducing fire risk in city buildings. Many of the provisions of these various acts are still in force today, incorporated in the Building Regulations in force at the present time.

Perhaps the most important legislation was that known as the Building Act of the 14th of George III (1774). It was this Act which empowered the Lord Mayor and Aldermen of the Justices of the Peace in quarter sessions to appoint district surveyors to ensure that building works were carried out according to the provisions of the Act. It was also made incumbent on the builder or owner of a property, prior to construction, to give twenty-four hours' notice of his intention to the district surveyor and, at the same time, describe the works to be erected or altered. In addition, three months' notice was to be given to the owners or occupiers of properties where the party walls were to be altered or repaired. It was directed that should the premises be empty the notice required under the Act should be stuck upon the front door or on the front wall of the property.

Houses were divided into seven rates or groups, and for each group the minimum permitted thickness for external or party walls was laid down. Groups were not designated as they are today by usage, but by ground floor area, this being described, not in square feet, but by units of 100 square feet (9·29 m²), or squares. Thus a dwelling house which exceeded nine squares of building (900 ft² or 83·6 m²) on the ground floor including internal and external walls was deemed to be a first rate building. Dwelling houses which exceeded three and one-half squares but were not more than five were deemed third rate housing. District surveyors were, under the Act, to be paid a salary, and also in the Act fees were laid down for the various approvals, which varied from £3·50 for a new first rate building to 25p for the approval of every alteration or addition to a seventh rate building.

Standards of building and precautions against the spread of fire were again made more stringent. Party walls had to be carried up at least 457 mm (18 in) above the roof slope and where dormers were placed within 1.22 m (4 ft) of a party wall the parapet had to be carried up to the full height of the dormer. Where openings were permitted in party walls, these had to be protected by an iron door of a minimum thickness of 6 mm ($\frac{1}{4}$ in). Hearths had to be built of brick or stone, extend at least 457 mm (18 in) from the face of the breast and at least 305 mm (12 in) on either side.

The minimum thickness of walls in buildings forming divisions between areas in different ownership were laid down in a manner similar to the compartmenting requirements in the present Regulations. It is interesting to note that specific exemption from the provisions was given in the Act in connection with the Inns of Court. The full details of the requirements make interesting reading, couched as they are in language familiar and in use today.

Further provisions were made in the Act of 1844. This amended the previous designation of groups, these being abandoned in favour of classes of occupation, which were then three in number: dwelling houses, warehouses, and public buildings. In this Act all halls, corridors and staircases in public buildings were to be fire-proof and no new building could be undertaken without the approval of a Central Office of Metropolitan Buildings which was to provide a registrar and two referees, whose responsibility it was to decide what was 'good, sound, fire-proof, fit, proper and sufficient' in relation to the requirements and provisions of this Act.

Although the problems of fire-fighting are a separate issue, they have become more closely connected with building as the availability of water supply and escape from burning buildings became matters of public concern. It was the acute shortage of water due to a long drought, lack of pressure in old timber mains, and carelessness in the conservation of existing supplies which led to the rapid spread of the fire in London in 1666.

In the years which followed, more attention was given to the provision of water supplies in cities, not only for drinking purposes, but also for fire-fighting. In this latter connection it is interesting to learn that when water was drawn from wooden water mains a suction pipe was inserted in a hole bored in a timber and provided with a fire plug when not in use. These were placed at prescribed intervals along the main, and after the Water Works Clauses Act of 1847 these fire plugs had to be maintained by the Water Companies. The passing of the Public Health Act of 1875 compelled all urban authorities throughout England to act in a similar manner.

The London Building Act of 1894 limited the height of the external walls of any building to 24 m (80 ft) with the addition of two further storeys in the roof space. This was the maximum height which could be reached by escape ladders and water jets from the hoses. The Amendment Act of 1905 gave a list of materials which could be accepted as fire-resisting but did not give guidance on the testing of the resistance of material. After the early efforts of the British Fire Prevention Committee, formed in 1897, the Royal Institute of British Architects requested the forerunner of the British Standards Institute to determine definitions of terms and provide a specification for tests. The Fire Offices Committee, in the meantime, had operated a testing station at Manchester and in collaboration with the Building Research Station had opened a new Fire Testing Station at Elstree. As a result it was possible to suggest fire-grading for many materials, which formed the basis for the revised bye-laws and Building Regulations now in force.

It can be seen, therefore; that the present regulations which govern the planning and construction of modern buildings have developed over many centuries and are the product of much experience and wide knowledge in the practical use to which we put our buildings.

CHAPTER TWENTY-TWO

Water Supplies

Two major problems of settled man are the provision of water supplies and the removal of human waste. The use of pottery vessels and human muscle power to transport water from lake or stream is still used by many primitive cultures today, but it was early discovered that by tapping a water source at a higher level, water could be brought by gravity in open channels. To prevent loss, these channels had to be lined with an impervious material and, where the water was liable to contamination, the necessity for a covering was evident.

Early examples of gravity fed channels can be found today in Persia, where they follow the general contours of the country-side in their passage to the villages. The Romans, however, in their direct and characteristic fashion, overcame the problems of contour and elevation by erecting vast systems of elevated aqueducts, which, by incorporating a stone lined channel, carried large quantities of water from source to the city fountains. Many examples abound in the countryside around Rome, but the best known example is perhaps the Pont du Gard near Nîmes in Southern France.

The Roman writer Vitruvius made important observations on the provision of aqueducts. In his books he recommended that the grading should not be less than 6 mm in 30 m ($\frac{1}{4}$ in in

100 ft) and that the channel should be arched over for protection against contamination. He also recommended holding reservoirs at about 5 km (3 mile) intervals along the line to maintain supplies against accidental or deliberate damage to the aqueduct. He discussed the dangers in the use of lead pipes for drinking water and advised the use of earthenware pipes. His book described the various methods of raising water from one level to another, by buckets and endless chain, water wheels, a water screw enclosed in a wood tube and bronze double cylinder water pumps. It is obvious from his book that water engineering in Roman times had reached almost the stage of an exact science.

While open channels for the movement of water remained the most important method, and we must remember that open aqueducts such as the New River supplied London with most of its water until the 19th cent., other methods were devised from the earliest times. Pottery pipes date from about 3000 BC and were excavated at the Palace of Minos at Knossos in Crete. These pipes, about 762 mm (30 in) long and tapering from 102-150 mm (4-6 in) in diameter were provided with a collar to allow the ends of the pipes to bear against each other. Pottery pipes of 1300 BC from Hazar in Palestine were provided with holes to enable branch pipes to be inserted, and pipes found in Greece were provided with well formed sockets. By Roman times, pipes were almost uniformly produced with parallel sides and sockets similar to modern drainage goods. These pipes may be seen at the Roman Palace of Fishbourne, near Chichester.

Methods of laying pipes varied. At Pergamon in Greece, a pipeline was discovered supported at intervals on stone blocks perforated for the pipes. Pipes were sometimes bedded in concrete, not only to provide support but also to seal the joints. An underground aqueduct at Lincoln was laid in this manner. Vitruvius recommended that the jointing of pipes should be carried out with a mixture of quicklime and oil.

Metals have been used for waterpipes from very early times. A pipeline dating from about 2500 BC and over 366 m (1200 ft) long was discovered at Abusir in Egypt. The pipe was about 88 mm (3 in) in diameter, formed from 1·5 mm ($\frac{1}{16}$ in) thick

copper sheet bent to form a tube and with a folded seam joint.
The pipe was set in mortar in a grooved stone both to provide
support and prevent leakage. Lead was used in great quantities
by the Romans for pipes, usually in lengths of about 3 m (10 ft)
and up to 274 mm (9 in) in diameter. The sheet lead was
formed over a mandril with a soldered lapped joint. Other
materials were used, rectangular blocks of stone bored to form
a conduit with occasionally spigot and socketted ends. Wooden
conduits, such as the one found on the site of the Bank of
England in London and to be seen in the London Museum,
were used. This conduit consisted of lengths of oak 197 mm (7$\frac{3}{4}$
in) × 115 mm (4$\frac{1}{2}$ in) bored 45 mm (1$\frac{3}{4}$ in) and joined by iron

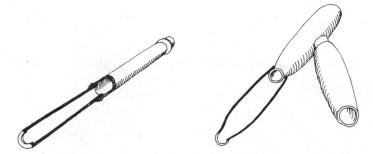

Fig. 126.   Water pipe from the palace of Knossos, Crete. 3000 BC

collars to enclose the channel. In Italy the Romans used pine,
pitch pine and alder for this work, but the availability of oak in
Britain caused them to use this local material. The abundance
of elm in mediaeval England, coupled with the priority in the
use of oak in building and boatbuilding, was responsible for
the popularity of the former material for water pipes. These
elm pipes remained in constant use until prohibited in the 19th
cent. Portions of lead piping still exist conducting natural hot
water to the bath house at Aquae Sulis, now known as the City
of Bath in Somerset. After the collapse of the Roman Empire,
it was not until late Norman times that water was again piped
to buildings, and then only to those of some importance such
as Christ Church Priory at Canterbury.

London received its first water supply in 1237 when the city was granted the springs and water of an estate at Tyburn by Gilbert de Sandford. The water was distributed through pipes, formed from elm trunks, to conduits or public fountains. From these, water could be drawn as required by the population. Better class families hired the services of water carriers and by 1496 these latter had formed themselves into a company of Tankard Bearers. Although private houses depended on these methods, monasteries frequently had a more or less elaborate system of water supply laid on to various parts of the buildings. This water was often brought from a considerable

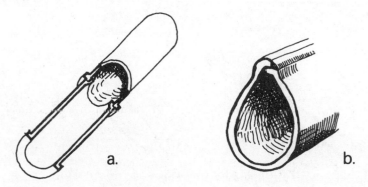

Fig. 128.   Roman water pipes   a. pottery   b. lead

distance. The monks at Canterbury, about 1150, installed an elaborate supply utilising a series of settling tanks to supply a raised cistern which fed a washing bowl below. An even more extensive installation was provided at Durham.

Where water could not be provided by a conduit, the source of supply would be a well. Whereas wells in Roman times were often constructed of baulks of oak set in pairs to alternate sides of the square shaft, such as the example which can be seen at Fishbourne, mediaeval wells were usually lined, at least for most of their depth, with regular masonry. Usually the water was drawn by means of a bucket and windlass, although the dipping beam and counter weight was in use at Canterbury.

Towards the end of the 14th cent. pumps were beginning to come into use. In 1504 Richard Lyncoln made three pumps and 19 m (62 ft) of pipe to convey water from the well to the wash house at Croydon.

The oldest municipal water supply is at Plymouth, where in 1591 an open channel or leat was constructed from a weir on the River Meavy to the City for watering ships and to provide protection against fire. Some of the water was diverted for public use, and some was also piped to private houses. Otherwise, few houses before 1600 were provided with piped water.

In 1606 the Corporation of the City of London was empowered to construct the New River to convey fresh water from the Chadwell Spring near Ware in Hertfordshire. The construction took four years, and the water was brought to Clerkenwell in London and is still part of the Metropolitan Water Board supply today.

As we have seen, wooden pipes continued for main distribution up to the beginning of the 19th cent., but due to considerable leakage water companies did not undertake to supply water above ground level because of insufficient pressure. London was divided into districts, each with an overseer, known as 'turn-cock', who regulated supplies to the cisterns used for storage. Although each district was supposed to receive a supply on three or at least two days a week, complaints were rife that often a week passed without districts receiving any water. It was generally found that the most effective method of obtaining a reasonable supply was to bribe the turn-cock. The problem was not generally solved until the introduction of iron pipes and steam pumping. This allowed the delivery pressure to be increased to ensure a constant supply which could reach the top floors of houses.

Many diseases, such as cholera and typhoid, are directly spread by polluted water supply. The earliest attempt to filter water supplies was probably made by the Glasgow Water Works Company in 1806, and in London in 1829 by the Chelsea Water Works Company. The intakes for supply were chiefly still, so far as London was concerned, in the tidal reaches of the Thames, fouled with sewage. In 1850, a Board of Health report recommended that supply for London should no longer be drawn from the tidal river and this was confirmed in the Metropolitan Water Act of 1852. This laid down that water was to·be drawn above Teddington Lock, and all reservoirs in the Metropolis were to be covered, or the water filtered. In addition all water from wells was to·be filtered. Finally the Royal Commissions of 1892 and 1897 recommended that the Water Companies should be acquired and managed by a single authority and in 1902 the Metropolitan Water Board came into existence.

Chlorination of water supplies was introduced at Maidstone in Kent in 1907 and there are now few public water supplies which are not treated in this way.

PART FIVE

The Rise of Professionalism

CHAPTER TWENTY-THREE

The Professional Designer

The training, methods and status of the building designer have varied greatly throughout the course of Western civilisation. From the first it was necessary to have some controlling hand in the layout and organisation of building works. The degree of skill required, and the philosophy behind the aesthetic principles and structural techniques employed varied greatly from one era to the next.

Four outstanding periods of history claim our attention, those of the ancient classical world of Greece and Rome, the Gothic of mediaeval times and the Renaissance. During these periods designers emerged, not only leaders of their profession but also in some cases teachers, whose writings served not only to instruct their contemporaries but also to provide us with detailed knowledge of the methods employed in their work. A brief study of the career of one designer in each of these four periods of activity provides an introduction to the emergence of the professional architect during the last 200 years of our civilisation.

The function of the Greek architect differed greatly from that of his modern counterpart. The English term is derived directly from the Greek word 'Architekton' meaning 'builder in chief'. As the normal plan of a 5th cent. BC temple was dictated by established tradition, the architect was free to realise this plan only in terms of Doric or Ionic columnar styles or orders. Later, a Corinthian variant of Ionic was added. It was not the architect's work to devise new and novel plans, or develop new ideas of construction and decoration. His role was to provide for the skilled marble masons schedules of measurement and full size patterns so that the vast number of identical stones could be prepared, and then to carry out the supervision of the construction and finishing of the work in its correct sequence.

The names of few Greek architects have survived; most of these artists have long since ceased to be known. One whose name has survived was Kallikrates who was active as a designer about 450 BC. Among his works were the Temple of Poseidon at Sunion, the temple of Athena Nike and temple of the Athenians on Delos. He was the originator of the attached pilaster which enabled the designer to articulate a plain wall surface. It has been suggested that he was in some way responsible for improvements to the Ionic columnar base towards its present contoured classic section. Another Greek architect was Iktinos, who between 448 and 437 BC constructed the Parthenon in white Pentelic marble, a masterpiece of sculptural architecture. Plutarch, writing of these buildings 500 years later, says that 'in their freshness and vigour they look as if they had been built only yesterday – an enduring spirit of deathless life possesses them'. It is perhaps one of the great tragedies of our time that further information and details of the work of these men have not been preserved.

Mention has been made of Doric and Ionic orders. Strictly speaking an order is the column and superstructure unit of a temple colonade and, while this does not necessarily have to have a pedestal or a base, it does have to have entablature and the cornice above represents the eaves of the building.

The earliest written description of any of the orders was by Vitruvius. He was an architect of some importance in the reign of Augustus, and wrote the classic book *De Architura,* a treatise in ten books, the only work of its kind to have survived from this period. Vitruvius was not a man of great learning or ability. His fame rests on the preservation of what might be regarded as the Roman code of building practice. In his third and fourth books he describes three of the orders, Ionic, Doric and Corinthian, and gives some notes on a fourth, Tuscan. In his first book, however, he gives his views on the education of the architect. Some of these seem curious to us today but many are still pertinent.

He says: 'The architect should be equipped with knowledge and understanding of many different branches of learning, because he is required to judge the quality of artistic work. Architects who have manual skill and dexterity without scholarship are not able to reach the professional heights which their position would warrant while those with scholarship and no practical skill hunt the shadow not the substance. Those who have a thorough knowledge of both practice and theory are in a position to obtain and wield authority.' The importance of Vitruvius is that his work was available to the architects of the 15th cent. – men like Alberti, who described the orders and added a fifth, the Composite, and Andrea Palladio, who was destined to provide a further impetus to classical design.

Andrea Palladio ranks among not only the most famous but also the most influential architects of all time. He was born in Padua in 1508 of humble parentage, and grew up in Vicenza. Originally trained as a sculptor and stonemason, he was taken to Rome in 1541 by his patron, Count Trissino, and there turned to the study of ancient classical building. He travelled widely in Italy, but spent most of his life in Vicenza, where he designed churches, town and country houses, and also many public buildings and bridges in Venice. Most of these structures were built of brick faced with stucco and have deteriorated with the passage of time. Among his best works were San Giorgio Maggiore in Venice and the Villa Capra near Vicenza. He died in 1580.

To what does Palladio owe his fame and influence? He insisted on the use of rules in design, firm in the belief that there is a correct and right way to design. He is the only architect after whom an architectural style is named – Palladian. His buildings were copied and imitated in their plans, elevations and details (Mereworth Castle in Kent by Colen Campbell; Chiswick House in London by Lord Burlington). His *Four Books of Architecture,* published in Venice in 1570, were later republished many times. The 17th cent. English architect Inigo Jones, during his visit to Italy in 1614, not only acquired a number of Palladio's drawings but also a copy of his book which, heavily annotated, is now preserved at Worcester College, Oxford. Amongst other buildings, the Banqueting House at Whitehall, 1619–22, and the Queen's House at Greenwich, 1616–35, result from Palladio's influence on Inigo Jones.

After the first English translation of Palladio's book was published in 1715 in London by the Italian architect, Leoni, the Palladian style became paramount in England, furthered

Fig. 129.   The Queen's House,
            Greenwich. 1619–35.
            Architect: Inigo Jones

by the efforts of Lord Burlington, an art patron of great influence and wealth.

The development of what may be termed the professional side of the building industry in England has been shown to have stemmed from the craft guilds of the Middle Ages. From these guilds came the development of the master mason and carpenter, selected for their organising, practical and aesthetic abilities to carry out the design and erection of buildings. These were the forerunners of the architectural profession as we know it today. Whereas Allan of Walsingham, Sacrist of Ely Cathedral, was responsible for the erection of the New Lantern, and Master Hugh Herland, master carpenter, is credited with the design of the great hammerbeam roof over Westminster Hall, in the main the rôle of designer was assumed by the master mason who was in general responsible for designing and controlling the whole of the work. The highest position to which these master masons could aspire in England was the post of Master of the King's Masons, which was held by a long series of great and distinguished designers. During the 14th cent. we find that this office was filled for more than forty years by Henry Yevele. His career is well documented and is a good example of the manner in which a prominent mason was employed in different works not only as a designer and constructor but also as a consultant and even more as an architect to supply the design from which the work was to be executed. This consultancy procedure is well illustrated by another mason – Stephen Lote – who was asked for advice in 1410 on repairs to Rochester Bridge; the records show that his fee for this work was 34p. On a lighter note, the Westminster Abbey accounts for 1479–80 show 25p spent 'on

Fig. 130. Villa at Chiswick for Lord
Burlington. 1725.
Architect: William Kent

rewards given, with a dinner, to three master masons for inspecting the old (part of the) church and the new, and their advice on repairs for the next year'.

During the 17th and 18th cents few English architects received any formal training and only the most fortunate received any tuition at all. It is likely that less than half had been apprenticed to an office and of the rest some had been trained by their fathers and a few, such as Vardy, Payne and perhaps Hawksmoor, were pupils or assistants in architects' offices. Many of the rest came into architecture from the position of Clerk of Works, mason or carpenter, while others approached it from painting or sculpture. Although training in an architect's office and travel in Italy were desirable, they were by no means essential and any ambitious craftsman who took the trouble to 'get up' the orders found a successful career open to him.

Recruitment, then, in the 18th cent. came from the following sources:

1. from builders, as apprentice to master-tradesmen (carpenter and mason to architect)
2. through pupillage, as clerk or craftsman in an architect's office
3. as an amateur by the acceptance of a high post in the Office of Works
4. from another occupation in the field of the arts.

The first method served those from the artisan class; the transition from builder to architect was easy when there was no great divergence in style, and we know that several builders became successful architects.

Pupillage does not appear to have existed before 1750, but several 18th cent. architects began as clerks in the offices of the Royal Works. Coming from every level of society, they were, with the exception of those in the Office of Works, invariably dependent on society patronage. The architect may have had some pupillage in an office, travelled in Italy, and maybe studied Palladio's works at first hand. On the other hand, he may have taken to architecture after much experience as a master craftsman, learning the principles of design from one of the published folios or Gibbs's handbook. As a knowledge of architectural principles was part of the education of a gentleman, there was a common bond of knowledge and agreement between the architect and the client.

The only institute at the beginning of the 19th cent. where pupils could receive instruction, additional to that received in their offices, was the Royal Academy of Art. Lectures were given and a library was available, but the inadequacy of the facilities led to discussions to found an association specifically for the architectural profession. In 1831 the Architectural Society was founded for the 'advancement and diffusion of architectural knowledge', the qualifications being a minimum of five years' study in an architect's office. Its main concern was with education and it remained little more than a library for students until it was merged with the Institute of British Architects in 1842.

By 1832 architects were finding conditions, between economic pressures on the one hand and competition from the unscrupulous on the other, intolerable without some means of defining their position and public image. Two years later the Institute was founded, and this gave the profession the example and stimulus needed. Societies modelled on the Institute sprang up in the provinces and many of them are still very much in existence today. In 1840 the Institute received Royal patronage and became the Royal Institute of British Architects; but it led a precarious existence, having in 1850 only twenty-eight Fellows, seven Honorary Fellows, seven Associates and two Honorary Associates. In 1871 the R.I.B.A. held its first general conference of architects to discuss, amongst other subjects, professional practice and education. Standardisation of fees and standards of professional conduct were the main concern in early years and, to the members, professional

Fig. 131.   The Customs house, Lancaster.
1764. Architect: Richard Gillow
(after Adam)

respectability was of more importance than professional technical qualifications.

In 1855, however, the Institute undertook the constitution of examinations in drawing ability and design, mathematics, physics, professional practice, materials, construction, and history and literature. In 1862 classes were started by the Architectural Association, whose members met regularly to criticise one another's designs, for members taking the R.I.B.A. examinations, and these have been continued and expanded to form the Architectural Association's School of Architecture as we know it today. Other Schools of Architecture have developed, attached to colleges and universities in most major cities.

Registration of architects was hotly opposed from within the Institute's ranks for many years after it was realised to be of prime necessity to the profession. Attempts to reconcile the position and promulgate the fusion of the R.I.B.A. and the Society of Architects took many years, and when the Institute's aim to close the profession became known, architect/surveyors woke to their danger and, opposing the R.I.B.A. plans, founded the I.A.A.S. and the F.A.A.S. in 1925 and 1926 respectively to guard the interests of unattached architects and surveyors against inequitable legislation.

In 1925 the R.I.B.A. absorbed the Society of Architects, and at length the Architects (Registration) Act 1931 was passed. This set up a separate body, the Architects' Registration Council of the United Kingdom, which maintained a register of architects. The Council set up a Board of Architectural Education, an Admission Committee and a Disciplinary Committee, these being the main responsibilities of the Council. In 1938 a further Architects (Registration) Act was passed which restricted the title of architect to those included in the register.

The whole concern of the profession since the 19th cent. has therefore been directed almost wholly towards the establishment and maintenance of the architectural profession by the guarantee of integrity and the maintenance and advancement of professional competence.

CHAPTER TWENTY-FOUR

The Measurer

The Royal Institute of Chartered Surveyors, which celebrated its centenary in 1968, has for many years worn two hats, those of land and quantity surveying. Originally, however, the main function of surveyors was the superintendence of property and this is the root from which the profession, with its many modern facets, stems today. The first important instance of the surveyor's work was the preparation of the Doomsday Survey: not a map or a measured document, but a product of enquiry, a valuation of property and rights, a document prepared by technically competent surveyors. This inventory was subsequently repeated many times in part for each manor, as each feudal lord needed to know the extent and condition of his estate. To avoid encroachment by neighbours this information was written down and, after a statute of 1276, according to a prescribed procedure. These surveys depended on memory and description to provide the information necessary to define boundaries. Only very rarely was a crude map used. If a dispute arose it was settled by custom and oath. It will be seen, therefore, that surveyors were legal or semi-legal in their activities, which led later to a domination of estate management by the legal profession.

In Roman times, however, land surveying was a relatively

Fig. 132.   Surveying in the Middle Ages

exact art. In chapter 5 of the eighth book of Vitruvius, written about AD 100 and dealing with the construction of conduits and aqueducts, practical advice is given on levelling. This account describes instruments then in use: the water-level, based on the physical law that water finds its own level in separate but interconnected vessels, and the 'Chorobates'. This instrument was based on a straight-edge, 6 m (20 ft) long, with legs and cross braces. When plumb-lines attached to the straight-edge coincided with the correct lines on the braces, the straight-edge was level. A groove 1·5 m (5 ft) long × 39 mm ($1\frac{1}{2}$ in) deep and 25 mm (1 in) wide on the upper face could be used as a water-level. By this instrument the fall of the ground over a distance of 6 m (20 ft) could be found. In addition, by using ranging rods suitably marked a method of levelling similar to the modern use of boning rods could be accomplished.

Fig. 133.   The surveyor at work

The Romans used ranging rods to set out their roads by alignment from one vantage point to another, roads which apart from deviations due to abnormal natural obstacles ran in straight lines from point to point. The use of a Chorobates was an advantage in preparing the gradients and cambers along the route. These methods used by the early surveyors were lost until the 15th cent. when their re-discovery coincided with a considerable expansion of surveying work.

In the 16th cent., the rise in population, the value of land and an active market led to a problem. The enclosure of land, often by fraudulent means, led to the necessity to define boundaries more accurately. It was no longer safe to rely on local knowledge to record property rights. Under such an impulse, the techniques of land surveying were revolutionised in the century from 1550. At the beginning of the period the only instrument known to the surveyor was a rod or line of variable local length. By 1650 surveyors were using chains of standard length, usually conforming to that devised by the astronomer Edmond Gunter early in the 17th cent., and they measured square areas in standard acres. In addition the surveyor had a plane table, a circumferentor and a theodolite, complete with a compass to fix position. As these latter instruments were of limited accuracy, the surveyor also had at his command a body of geometrical and trigonometrical information which made possible triangulation and other calculations, and the determination of angles solely by chain. Provided with scales on the vernier principle and field books, he was well prepared to measure, observe and plot with reasonable accuracy.

Much 17th cent. development in land and estates produced demands for the services of surveyors. Having mastered the art of measuring and plotting, surveyors moved into other fields of estate work such as valuing, property transfer and improvements. The 18th cent. development of estate management was spurred on by the invention of new agricultural techniques and the drive for improvement. It was at this time that surveyors began to replace lawyers as stewards of property and estates. At the same time, there began a movement for the conventional land surveyor to establish a foothold in land agency, and from then on the movement into estate management was inevitable. The further enclosure of land in the 18th cent. brought an enormous amount of work to the land surveyor. As this boom in business moved to its peak, the Ordnance Survey was launched which was to sound the death knell of chain surveying for all time. The need for accurate maps of Britain was brought home to the authorities during the campaigns of the 1745 rebellion. Due to delays and interruptions, little was attempted until 1787 when a base line was laid down on Hounslow Heath for the triangulation to determine the relative positions of the observatories in Britain and France. From this base line the principal triangulation of South East England was prepared under the supervision of officers of the Royal Artillery.

The first Ordnance Survey one inch map, of Essex, was published in 1805. County maps had been produced for many years before this by cartographers such as John Speed and Robert Morden. Many of these maps were beautifully engraved and illustrated.

The evolution of canals, of which 3000 miles had been constructed in Britain by 1830, did not bring much work to the surveyor. Nor did the expansion of railways during the 19th cent., apart from work in connection with land acquisition. The men who carried out this part of the work would today be termed civil engineers, and, like John Rennie, called themselves the 'principal engineers'. This work was usually performed by relatively junior staff employed for the purpose by the consultant engineer surveyor.

During the Middle Ages building projects or building organisations of any size employed a staff to make payments for wages and material and to keep accounts. Payments for parchment or paper for these 'rolls' of expenditure and for clerks' wages occur often in mediaeval accounts. Sometimes a single Clerk of Works was employed, often several. Sometimes the clerks specialised, for instance in mason's and carpenter's work, or for bricklayers and allied trades. These men were the forerunners of the quantity surveyor as we know him today, filling a vital position in the contracting organisation of most major building firms.

The roots of modern quantity surveying, however, go back into the 17th cent., and the Great Fire of London. The 'measure and value' system of settling payments was in use before this, but the vast amount of work necessary after 1666 gave an impetus to measuring and the employment of separate measurers. Architects were inundated with design work and had no time to spend measuring their own buildings. To enable accounts to be settled the work was given out to these new specialists. In 1685 tables were published for measuring materials and labour. The usual practice was for two measurers to be appointed, one for each party to the contract, and the system was obviously open to abuse. By 1770, however, it appears that something similar to present day practice of making pre-building estimates from drawings was coming into vogue, and Bills of Quantities began to be prepared. The establishment of the Barrack Office in 1793 to erect the large barracks required in the Napoleonic Wars gave impetus to this development. Measuring surveyors were appointed on a part-time basis to apply 'The London Mode of Measuring' to take measurements and adjust quantities.

In 1834, a group of old established land surveyors held the first meeting of the Land Surveyors' Club. These men regarded their prime function as valuers. They also had strong views on professional status and rigidly excluded from membership anyone who did not meet their strict and exclusive conditions of membership. They intended to advance the dignity and status of their profession, which they regarded as synonymous with private practice. However, the Club became too specialised in its estate and agricultural work, the times were against its advancement and it was closed in 1852.

In March 1868, the Institution of Surveyors was formed under the presidency of John Clutton and took a sublease on No. 12 Great George Street, Westminster, a building which the Institution occupies today and whose freehold they purchased in 1958. In the first few years members were recruited by persuasion, but the growing prestige of the Institution gained a recognition in the Census of 1881 where surveyors were ranked in the 'professional class'. By 1880, the Institution had reached the position when it was at a disadvantage without a Royal Charter, and its acquisition made membership attractive or at least advisable. The Institution was now the accredited representative of the profession and became morally responsible to the public for the standard of competence and integrity of its members. Education became a duty of the Institution, although its examinations had been resisted for some years. These were first held in 1881 just before the application for the Charter came before the Privy Council. Since that date, the style and content of examinations have been under continuous review to ensure that they are relevant to the present needs and activities of the profession.

The Institute of Quantity Surveyors came into being for a very different reason. Many quantity surveyors were employed in the offices of contracting organisations and as such were precluded from membership of the R.I.C.S. It was felt by many that some form of association would be of benefit not only to those so employed but also to the profession as a whole. This occurred in 1938 and in 1941 The Institute of Quantity Surveyors was incorporated, membership being gained by practice qualification. In 1942, however, the first steps towards closing this door to entry were taken with the institution of the first professional examination. In 1948 the Institute moved into No. 98 Gloucester Place, London, which it had acquired on a short lease and in 1952, at the sale of the Portman Estate, the freehold was purchased and the Institute could look forward to a continuing and secure future. In 1960 the 'Limited' was removed from the title of the Institute. Membership has made steady progress from the initial one hundred and six members in 1940 to over seven thousand today.

CHAPTER TWENTY-FIVE

The Builder

Changes in the methods of contracting for building works over the past six hundred years cannot be shown as a clear cut sequence of events. Changes which occurred took many years to spread to universal practice and, in some respects, mediaeval practices remain in use today in Scotland. This is not to say that Scottish practice is old-fashioned. Present trends are moving towards a system whereby a building manager, employed direct by the building owner in a similar manner to the architect and quantity surveyor, is responsible directly to the employer for organising a series of subcontracts for the work, a method universally employed in Georgian England for major works.

As we have seen before, building organisation in the Middle Ages utilised a system of craft contracting, organised and controlled by a master mason assisted by Clerks of Works and administrative assistants. This system worked well while building remained a comparatively simple operation, employing relatively few crafts. The designer had little difficulty in calculating the cost of the work, based on his own experience of similar work carried out in the past. When authorisation was given for the work to proceed, estimates were invited and received from master craftsmen either on a lump sum basis or a schedule of rates. During the 17th and 18th cents this method sometimes included a Bill of Quantities.

We are very fortunate that the records of the building of the Radcliffe Camera at Oxford have survived, giving us a clear insight into the methods of contracting in the 18th cent. The design of the building was successively in the hands of several architects, the eventual design being the work of James Gibb (1682–1754). The schedules for the various craft works were prepared by Thomas Jersey, Clerk of Works, and these were priced by selected master craftsmen. The separate tenders were then compared by Jersey who prepared a cost comparison which was sent to the client. The lowest tender was then accepted. When the work was complete, it was measured and priced out at the rates in the original schedule by Thomas Jersey who agreed the figures with the master craftsman concerned, who was accordingly paid the monies due to him. This was the standard practice adopted for Georgian building contracts.

The procedure described above for measuring completed work was probably used on larger contracts until the start of the 19th cent., and even later. Clear lines of demarcation are, however, difficult to draw, as one or two firms of building contractors have been carrying out work since the middle of the 18th cent. in a manner similar to present day practice. On the whole, though, it was not until building works began to reach a state of some complexity at the end of the 18th cent. that changes in traditional organisation were necessary. General supervision of the various trades and their organisation began to be placed in the hands of one or other of the master craftsmen, who, for his pains, charged a fee, or took a profit on the total value of the subcontract works. In some cases, of course, the work was carried out at a loss, and this contractual risk became accepted and efforts to cover this unpleasant possibility were made in subsequent contracts. This form of contracting procedure was hastened by political events in Europe which required considerable building activity in the public sector to support the country's effort in the Napoleonic wars, and also by the rapid growth of industrialisation with its need for vast expenditure on mills, factories and workers' houses.

This trend was accelerated in Victorian times when the emergence of the general building contractor, employing men of every trade throughout the year and carrying out the whole work involved in the contract, is associated with the development of the growing practice of 'contracting in gross', that is, for one builder to offer to erect the whole of the building at a predetermined price. Between 1800 and 1830 there was much discussion on the subject of the best method of contracting for building works. The matter was raised in Parliament and was the subject of Select Committees. The Napoleonic Wars had resulted in a great spate of building of barracks, hospitals and prisons. With the return of peace, economies were demanded in the sphere of public expenditure and excess over estimated cost did not find favour with an economy-minded public. A growing number of protagonists contended that contracting in gross reduced the public liability to excessive expenditure, a point of view diametrically opposed by the traditionalists, who maintained that any advantage of the new system was illusory.

The gross contract for a single trade was, as we have seen, common in the Middle Ages. The chief argument against the practice was that a contractor making too low an offer might either suffer a financial loss or cut the work to ensure his profit. His loss might bring about bankruptcy, and as early as 1356 the London Regulations for the Trade of Masons required that each contractor in gross should provide sureties from men of his own trade. The risk of bankruptcy was increased in the 19th cent. by the development of competitive tendering. Tenders were called for by advertisement or by selective nomination. The great disadvantage of the new system was that the architect did not necessarily have personal knowledge of the tradesmen he was employing, and often tradesmen quoted uneconomical prices to secure work as an advertisement. This was common to work on royal property. Limited competition from chosen tradesmen was a safeguard against these charges and in the 1820s this system was used by the Office of Works for all its major building projects. At the same time, the Office was firmly against the system of contracting in gross.

Contracting in gross was not an innovation of the 19th cent. Some of the principal building trades – bricklayers and carpenters – frequently undertook contracts in gross in the 18th cent. There were obvious risks, and when short of ready money the subcontractor was often paid in kind, and in any case allowed the principal a discount. An architect, Robert Morris, asserted that financial pressure on the subcontractor led to the employment of inferior workmen or materials. The supply of materials by the contractor he regarded as a serious menace, as this deprived the subcontractor of his advantage of competitive purchasing, leaving him with only the value of his labour. The situation is paralleled today by what is now known as 'the lump'. It must be remembered, however, that many carpenters were also timber merchants, bricklayers made their own bricks and quarry masters had long been practical builders. Whatever the problems, it was apparent that the larger the contractor's business, the larger the capital at his disposal and the more able he would be to ride a contractual loss.

John Nash, the architect, was one of the early advocates of contracting in gross. He maintained that the new system would require the architect to prepare a detailed specification setting down all materials and labours. Not only would this ensure that the architect carried out his work properly in the first instance but also that the supervision would be more effective, the Clerk of Works having a technical document at his disposal stipulating the exact technicalities of the work. A further advantage, Nash maintained, was that damage by a tradesman in carrying out his particular work would be put right at the contractor's expense and not, as so often happened, find its way onto the bill of extras.

It is obvious that the prime factor for the success of gross tendering lies in the quality of the specification. With a loosely worded and constructed document, contractors were ready to quote a low offer, trusting to their ability to twist matters to their own advantage in extras. It was obviously easiest for architects initially to prepare specifications for simple buildings, and the earliest contracts in gross of any size were those placed by Government Departments for barracks and hospitals and by commercial concerns for warehouses. From 1805 the Barrack Department employed a form of contract requiring work to be carried out for a specified sum according to plans and specifications signed by both parties. Payments were to be made by instalments as the work proceeded, extras paid at a fair valuation and a bond or surety provided by the contractor. This procedure closely resembles present day practice for small works contracts.

The Victorian method of tendering for building work was different from present day practice. When the architect had completed his drawings and prepared his specification he would either advertise for or invite contractors to tender for the work. When the list was complete, it was the custom for the contractors to meet to appoint a surveyor, on their behalf, to take out quantities for their use. Sometimes two surveyors were appointed, one to serve the interests of the contractors and one the employer, to take out quantities together and check the result as the work proceeded. Each contractor, being provided with a copy of the Bill of Quantities, prepared his tender, which was delivered to the architect. It was usual for the tenders to be opened in the presence of both the contractors and the employer. The preparation of the contract, supervision, adjustment for variations and certification of cost followed in general principles present day practice. It is interesting to note that surveyors were, even then, allowed a commission of two and one-half per cent on the cost of the work, but were responsible to the contractors for any omissions to the quantities.

In 1834, a small group of the leading master builders in London formed The Builders' Society to 'uphold and promote reputable standards of building through friendly intercourse, the useful exchange of information and greater uniformity and respectability in building'. Members used the meetings to discuss not only professional matters but also labour relations and business policy. In 1859, the Central Association of Master Builders was formed, the forerunner of the London Master Builders' Association, now the London Region of the National Federation of Building Trades Employers.

In 1884, the Builders' Society was incorporated and became the Institute of Builders, membership remaining limited to principals of construction firms in the London area. The main preoccupation of the Institute until the end of the century was with building legislation then being formulated and, in addition to the Building Acts, the Standard Form of Building Contract which was eventually issued in 1909. After the First World War the early editions of the Standard Method of Measurement of Building Works occupied much of the Institute's time. Interest in education began in the 1920s and in 1924 the Institute offered its own examination for the Licentiate Diploma, following this in 1927 with the Associate Diploma Examination. During the years 1959 to 1964 the examination and membership structure was revised and the Associate class was given corporate status in the Institute. In 1965 the name was changed to The Institute of Building.

The Federation of Master Builders was originally formed in London in 1941 to assist small building employers who had been brought in by the Ministry of Works for the purpose of carrying out urgent and primary war damage repairs. There was little cohesion among employers and virtually no universal form of administrative regulation or standard contract. For these reasons builders found it extremely difficult, when dealing with local authorities, to regulate the working conditions for which payment was to be made. The Federation, originally known as the Federation of Greater London Master Builders, was in 1943 constituted on a national basis and received its present title. In 1951, the Master Builders' Federation was formed to take over the industrial negotiation side of its work.

CHAPTER TWENTY-SIX

The Engineer

Engineering first came to the fore in the years between 300 BC and AD 500. This was the period in which Rome dominated the Western civilised world. During this time, only one novel method of production appeared, that of blowing glass. Otherwise, the technological advances of previous ages were applied and developed.

Engineering was the only aspect of technology in which an intelligent man could participate, and the most important centre for its development was Alexandria. In this city a Greek general, Ptolemy (or Ptolemaeus) established the 'Museum', essentially an institute for teaching and study, incorporating a library which became world famous. To this centre came scholars to learn and teach, among them a Greek called Hero (or Heron), known to history as the compiler of the first engineering textbook. Most inventions developed at Alexandria depended on the simple principles of the syphon, the spring, the screw and other such elementary devices. Many of these are described in the writings of Vitruvius, e.g. Hero's water clock and the fire-engine devised by Ctesibius. Perhaps the greatest engineer of the period was Archimedes, a native of Syracuse, who invented a screw pump for lifting water, as well as systems of compound pulleys for lifting heavy weights. This last invention was of great practical application in building and cranes were used by the Romans based on the principles developed by Archimedes.

The greatest contribution of Rome to the advance of technology was the ability to absorb ideas from other countries, and to rationalise and improve on the original concept. For example, while the Greeks considered mathematics an interesting mental and aesthetic exercise, the Romans used its methods to survey the route for their roads and aqueducts. One of the most useful artifacts developed during the Empire was the water-mill, usually set horizontally and operated by a jet of water in the manner of a turbine. A different kind of water-

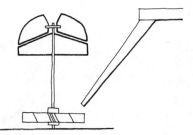

Fig. 134.   Horizontal water mill

mill, the under-shot type, was described by Vitruvius, and in the 5th cent. AD a further type, the over-shot wheel, was developed. This type continued in use in England and other countries for grinding corn until the present century. This mechanical device was also used by 18th cent. engineers such as Arkwright to provide power for mills built in the early days of the Industrial Revolution. The Romans, however, could not appreciate the use of this vast source of energy apart from that of milling and continued to use muscle power alone to smelt and forge iron for the Imperial requirements.

After the collapse of the Roman Empire, the philosophical application of pure engineering was extinguished until the impetus of mediaeval expansion necessitated the provision of power to aid man in the vast building projects for which this period is known. The cathedral builders were active participants in Europe's first industrial revolution. Natural sources of power are a basis of a civilisation's industrial progress and three principal sources open to the Middle Ages were water, wind and horse power.

As we have seen, the water-mill was extensively developed in Roman times, and its use spread over the whole of the empire. Problems were experienced, however, in regulating the supply of water to provide for continuous operations, and the expansion of water-mills in the Middle Ages might well be found to be due in part to changes in climatic conditions which

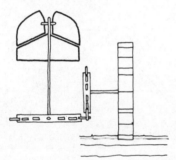

Fig. 135. Vitruvius's undershot
        water mill

Fig. 136. Overshot water mill

provided regular annual floods in the Northern rivers, together with a greatly expanding population. The invention of the camshaft, allowing a rotary movement to be changed to reciprocal movement, enabled hydraulic power not only to grind corn but also to forge iron and assist in the production of paper. In a few instances tidal mills were constructed in suitable estuaries. With developments in the construction of cogwheels, the problems of constructing windmills were overcome and from the 12th cent. vast numbers of these elegant constructions were erected throughout the countryside.

The horse was a source of considerable power for the Middle Ages, and cathedral workshops made use of it not only in the movement of material but also in the lifting of heavy weights. This was mainly due to improvements in the design of harness, and the protection of horse's hooves by having them shod with iron.

The cathedral builders were not only progressive but also inventive. Most of the craft trades developed simultaneously, and each helped the other in their work. Blacksmiths produced stronger steel tools, which, by cutting harder stones, enabled the carvers to execute more delicate work. Besides tools, tie-rods for preventing walls from spreading and iron chains for reinforcing walls were manufactured in the smithies of the Middle Ages. Carpenters also perfected tools which permitted improvements in carpentry and scaffolding. Mediaeval carpenters developed techniques which enabled them to shore up

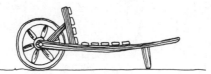

Fig. 137.   Mediaeval wheelbarrow

existing structures not only for under-pinning the foundations, but to enable modifications to be made to first floors. The wheelbarrow was developed and made its appearance in the 13th cent., allowing one man to do the work of two.

These advances cannot be ascribed to any single person or community. The age was one of great inventiveness and as much of the application was for the use of the universal church, the spread of ideas was no doubt aided by travelling churchmen, which probably accounts for the rapidity with which new ideas spread throughout Europe. There was, however, no organisation for the training of engineers, or the practical dissemination of ideas. Most of the practitioners were either practical masons or members of the religious orders immediately concerned with construction.

An exception to this rule was Leonardo Da Vinci, who, while

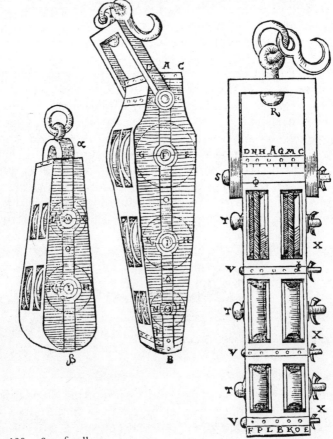

Fig. 138.   Set of pulleys

living in Milan in the latter part of the 15th cent., produced a
vast number of fine drawings of engineering subjects. Many of
these drawings were for military use: cannon, catapults and
winches. One of the finest of these, dating from about 1500,
showed a machine actuated by a turbine for shaping and
tapering bars of iron for building large cannon. Leonardo also
studied the problem of continuous artificial flight, including
proposals for a form of helicopter, later turning his thoughts
to soaring with devices for stabilising the machine in flight.
Although no building exists which can be credited to

Fig. 139. Levering machinery

Leonardo many drawings of architectural subjects, especially domes, survive. In addition, plans and drawings exist for a model city, and these show a most imaginative, practical and modern grasp of the problem.

Engineering techniques continued to develop throughout Europe, especially in England throughout the 16th and 17th cents, culminating in the immense expansion of the Industrial Revolution. This is described in some detail in the separate section dealing with the Industrial Revolution in England during the 17th, 18th and 19th cents. It was well into the 19th cent. before the profession of engineer became respectable. The raising of any professional status comes from within, and consequently it is necessary for a body of individuals to exist to make rules and enforce their observance. With the Institution of Civil Engineers this occurred in 1771, when the Society of Engineers was founded. Engineers in those days, and there were few of them, were usually the sons of artisans. An exception was the first president of the Society, John Smeaton, the son of a well-to-do attorney. The original proposer of the Society spoke of the 'profession' and it is likely that Smeaton saw the possibility of advancing the social status of the engin-

Fig. 140.   Post-mill at Outwood, Surrey.
1665

eer. Unfortunately the Society did not flourish and after Smeaton resigned in 1792 it was dissolved by mutual consent.

For the next quarter of a century engineers were busy on the vast programme of work necessitated by the military and political problems of the war with France and they do not seem to have taken much interest in the formation of any professional association. However, in 1818 the foundations of the society which was to become the Institution of Civil Engineers were laid. A group of eight men used to meet in a coffee house in Fleet Street and, being engineers, they 'talked shop'. One, Henry Robinson Palmer, proposed that they should form themselves into a society. Palmer was well educated, a Fellow of the Royal Society, but apart from his work as Vice-President of the Society he originated, little is known of his professional career. The Institution made a poor start and it was clear that it lacked a driving force to direct its affairs. In 1820 Thomas Telford agreed to become President and in this man the driving force was found. Telford was 62 at the time, not only the leader of his profession but a man of energy and integrity. Within six years, membership increased from thirteen to one hundred and seventeen and the Institution occupied its own premises in Buckingham Street, London. Members thought it was time they had a charter, and in 1828 application was made to the Privy Council. In this document was included the perfect definition of civil engineering: 'Civil engineering is the art of directing the great sources of power in nature for the use and convenience of man.'

The Royal Charter was duly granted and remains with minor amendments the deed of incorporation of the Institution. Because of it, corporate members have the right to inscribe 'Chartered Civil Engineer' after their names, implying that they are qualified engineers and members of an institution deemed worthy to hold a charter under the royal hand.

We have seen that the work of civil engineers occupies a clearly defined field. Its materials are natural substances. It has little to do with manufacturing processes or machines. Mechanical engineers such as railway engineers, of whom George Stephenson was the doyen, were too busy in the early years of the 19th cent. to bother about professional status. When it was rumoured that Stephenson had had his application for membership of the Institution of Civil Engineers deferred until he could submit evidence of ability, the time and the climate were right for the formation of a society where mechanical engineers could meet to discuss their own particular problems and difficulties. Mechanics' Institutes and Mutual Improvement Societies were springing up and the spirit of learning was stirring. In 1846 the initial moves were made at a meeting at the Queen's Hotel, Birmingham, and in the following year the Institution was formally established and Stephenson was elected the first President.

The Institution of Mechanical Engineers remained a

Birmingham society for many years and membership remained fairly static at about two hundred. When it was decided to hold summer meetings in various other provincial cities, membership began to expand with a growing interest in the aims of the Institution. A meeting in London in 1873 was an unqualified success and in 1877, by a large majority vote, it was decided to move the headquarters to London. After occupying 10, Victoria Chambers, Westminster, for some years, the Institution purchased the present headquarters at Storey's Gate in 1899. Application for a Royal Charter had been made immediately the Institution moved to London but this was not successful. It was not until 1929 that a second application was made and this time the Royal Charter was granted in the following year.

Although the Institution of Electrical Engineers had a short-lived forerunner in the Electrical Society of London in 1831, the accepted birthday of the Institution is May 17th 1871 when a Society of Telegraph Engineers was formed. This is interesting as it shows that at this time the only widespread use of electricity was the telegraph. Nine years later, however, the addition of the words 'and Electricians' showed that the Society was enlarging its interests, and six years later all reference to the telegraph was dropped and 'Electrical Engineers' was adopted. This shows the rapid advance which the science and technology of electricity was making, especially in its application to modern life.

The first President was Sir William Siemens and in 1872 the membership was three hundred and fifty-two, rising in four years to six hundred and fifty. By 1887 it was clear that the Society must expand and it was decided to change the title to The Institution of Electrical Engineers to embrace all facets of electrical work; this change was confirmed in 1889. For the first thirty-eight years of its existence the Institution had no home of its own, but in 1910 the present headquarters in Savoy Place was acquired. As far back as 1873 the Society had thought about a Royal Charter, but when formal application was made in 1880 it was refused. A second application in 1921 was more successful.

The Institution of Structural Engineers began life as the Concrete Institute, a body founded in 1908 to bring together for discussion and common action persons interested in the study and development of reinforced concrete, at a time when this was receiving little attention from the other institutes. As early as 1850, the embedding of iron in concrete was tried experimentally, in France for the construction of tanks and in England for floors. By 1880 this system had attracted attention in Europe and North America, but little interest was shown in Britain until L. G. Mouchel opened a design office in London to exploit French patents in this form of construction. By the first decade of the 19th cent. there was a flood of patents for systems of concrete construction.

In 1906 the British Fire Prevention Committee appointed a special commission on concrete aggregates and the R.I.B.A. appointed a Reinforced Concrete Committee. The first report of this committee appeared in the R.I.B.A. Journal of 15 June 1907 and the Council of the newly formed Concrete Institution first met on 21 July 1908. It was agreed to incorporate the Institute as a technical society and this was effected on 22 February 1909. The Institution quickly became accepted as a consultative body, and in 1920 the first candidates sat for the Institution's examinations. In 1922 the Council voted for a change of name to the Institution of Structural Engineers and in March 1934 the Royal Charter was granted.

All four engineering institutions are vitally interested in education and conduct their own examinations to ensure a proper standard of ability, knowledge and training for all aspirants to corporate membership. Registration of professional engineers has been attempted on several occasions but the opposition from members has always defeated the attempt, as it has been felt that corporate membership of one of these institutions means a good deal more than 'registered engineer'.

CHAPTER TWENTY-SEVEN

The Craftsman

Fine building is a combination of the skills of designer and craftsman. These skills are complementary to each other and a period of history fortunate enough to possess both enjoys a quality in its buildings which becomes part of our national heritage.

The prominence of particular crafts in any one period of history is dependent on the available structural techniques and materials, and practice makes, if not for perfection, a reasonable substitute. The problem for the student of building history is to decide which comes first, craftsman or craftsmanship. It must be understood, however, that throughout the whole evolution of Western civilisation fine craftsmanship has always survived the efforts of man to change the course of events. Subsequently, when new fashions produce a reversion to previous craft practices, men can always be found both to carry out the work and to pass on their inherited craft knowledge.

Over the years the prominence of the various trades in the pattern of building has ebbed and flowed. Originally the structural trades were dominant, the mason and the carpenter, producing with their inherited craft skill structures of great beauty and ingenuity. Fashions changed, comfort became a desideratum, the decorative. arts flourished and became dominant in the design. Each of these periods produced men of genius in their particular skills. Unfortunately our knowledge of their careers is fragmentary, based largely on chance references in contemporary writings, accounts, parish registers. What has survived provides a fascinating insight into the training and general practice followed by these men and their contemporaries. All conformed to a well defined pattern of craft apprenticeship, all were educated far beyond the extent considered either necessary or desirable for the tradesman of the present day. All considered it essential to be conversant with the design techniques and vocabulary of their time, enabling them to interpret correctly the ideas and instructions of the designer and his patron, and all considered that travel and study were necessary adjuncts to the skills of the well trained craftsman. How pathetic, in this respect, are the syllabuses provided for craft training schemes today.

The names of certain men stand out in the roll of master craftsmen. Our knowledge of their lives and professional activities varies; some have risen by their own efforts, some have carried on and enhanced a family tradition. All have benefited from patronage. Three men, a mason, a smith and a plasterer have been selected from their ranks as representatives of the great craftsmen of the past, to illustrate in a fragmentary manner the changing world of the craftsman over a period of some four hundred years.

Henry Yevele was born about 1320, probably in Yeaveley in Derbyshire. His father, Roger, was a mason by trade and he may well have been employed on the building of the spires of Uttoxeter and Ashbourne churches then under construction. This part of the country formed a portion of the immense estates owned by the house of Lancaster and while the details of craft mysteries such as drawing, the use of square and compasses and practical skill in hewing and preparing stone would be learnt from his father, the neighbouring Priory of Tutbury would provide opportunity for acquiring a knowledge of geometry and Court French and Latin. Tomb carving was a local industry, utilising the locally produced alabaster, and it was probably here that Yevele learnt the skills of the carver which stood him in such good stead in later years. It is also likely that he was influenced by the new work at Lichfield Cathedral then coming to completion under the supervision of Master William Ramsey, the King's Master Mason.

Little is known of Yevele's life or work before the plague of 1348–9. That he survived may well indicate that he was working in the provinces. Many craftsmen died, among them William Ramsey and his brother John. New men were required to take their place. It must have been about this time that Yevele settled in London, for by 1356 he had become one of the principal free masons of the City. At this time, London had no regular guild and disputes arose between the hewers and the layers and setters. The first group included the master masons who were also the principal architects of the period. The quarrel reached a head in 1356 when the Mayor – Simon Francis – intervened in an attempt to remedy the troubles. Six representatives of each side were elected to represent their respective cases before the Mayor and Aldermen, and one of the hewers was Henry Yevele, now aged about thirty-five. The Articles which resulted regulated the work of masons within the City, together with rules for the employment of journeymen and apprentices.

Yevele's work in this connection stood him in good stead, for in 1358 he undertook to build certain walls, chimneys and staircases at Kennington Palace for Edward the Black Prince and a year later he became his mason. After the signing of the Peace of Bretigny, between England and France, Edward III was at liberty to proceed with the great works at Windsor Castle delayed by the war. Yevele was appointed 'Disposer of the King's Works in the Palace of Westminster and the Tower of London', in effect chief architect for the whole of Southern England. Under his control were the vast establishments at these two places with their counting houses and storerooms, tracing houses and glaziers' lodges.

One of the first projects for Yevele was the construction of Queenborough Castle on the Isle of Sheppey, started in 1361. This was designed purely as a garrison fortress, contained within a circular curtain wall and surrounded by a moat. The

255

whole conception of this castle was well in advance of its time and Yevele acquired a reputation for military engineering which brought him further work in later years. The Clerk of Works in charge of the building was William of Wykeham, later to become Bishop of Winchester.

During the next four years many commissions were carried out by Henry Yevele: the chantry chapel in the crypt of Canterbury Cathedral for the Black Prince, the Abbot's Hall at Westminster Abbey, the College of Chantry Priests at Cobham. By 1378 a succession of French raids on the South Coast caused considerable alarm and the Bailiffs of Canterbury erected the city wall and Western Gate; two years later Lord Cobham built the great gate to his manor house of Cooling near Gravesend,

Fig. 141. Gateway to Saltwood Castle, Kent. 1383

and three years later the Great Gatehouse at Saltwood Castle near Hythe was put in hand – all attributed to Henry Yevele.

Three other works engaged his attention at this time. The first was the Western extension of the nave of Westminster Abbey in keeping with and incorporating the five Eastern bays of the original design of the previous century. The second was the rebuilding of the nave of Canterbury Cathedral started by Archbishop Sudbury; and the third, perhaps the greatest work of all, Westminster Hall in collaboration with Hugh Herland, Master Carpenter. In addition Yevele was responsible for carving the tomb of Edward III at Westminster Abbey.

True to his generation, Henry Yevele continued to produce work of great quality well into his prime. It is probable that in his last years he was able to live the life of a country gentleman on his estates at Southwark, Wennington and Aveley in Essex or at Deptford and Greenwich. He died on 21 August 1400 and was buried in the tomb he had prepared in St Magnus, Southwark, inscribed with his name and the date of his death and the modest epitaph that he had been employed as freemason to three kings, Edward III, Richard II and Henry IV.

The first expression of a renaissance in English ironwork is difficult to separate from the last productions of the mediaeval smith, since the two moved forward in parallel for many years. Probably the best examples of the new art are to be found at St Saviour's, Dartmouth, where the wrought iron columns, 150 mm (6 in) in diameter, supporting the gallery are the largest forgings produced in Europe before the introduction of the steam hammer. A form of strapwork consisting of thin iron strip with the broad face on the vertical instead of the horizontal plane is a feature of work of this period, together with long plain or twisted bars terminating in flattened spikes with a similar C scroll welded on either side. Examples can be seen at Cirencester and Hereford. This style ran a considerable risk of being displaced by the great French smith Tijou who settled in this country but, instead of permanently displacing the national style, Tijou's rich and florid designs passed away and were absorbed leaving ironwork design, at the end of Queen Anne's reign, very much as it was under Charles I.

It is remarkable that so little is known of Jean Tijou, the great French smith who came to England after the Restoration, especially as he created a new school of ironwork in England and exerted an influence over his associates which was felt long after his death. Tijou was, it is presumed, one of the French Protestants who fled to Holland in 1685 after the revocation of the Edict of Nantes, and came to England after the accession of William and Mary. While workers in decorative iron contribute less to a building than joiners and plasterers, their balustrades and gates enhance the overall conception. The first record of Tijou's work is in 1689, when he was working at Hampton Court, through his bill for the six richly wrought iron vanes and a wrought iron balcony in 'finely wrought leaves and scrollwork'. He gained the notice of Queen Mary and in view of her absorbing interest in architecture and gardening, a brilliant career seemed to be his. He was commissioned to execute the garden screen for the circle of the fountain garden and for the three perfectly preserved gates filling the archways of the East front of the palace, at a cost of £1115·62. In addition, Tijou was commissioned to carry out work at Chatsworth and Burleigh, while his work for Wren at St Paul's was just commencing. At the height of his popularity his patron, Queen Mary, died and at once all work at Hampton Court was abandoned, until the destruction of Whitehall by fire in 1698 caused William to make Hampton Court his principal residence. Even so, the resumption of work did not mean new commissions for Tijou, only one of the outstanding works, that of the King's Staircase balustrade, being required from him. The design for the new railing required for the East front of the palace was prepared by the architect William Talman, who estimated the weight at nearly 46 tons and the cost at 2p per 453 g (1 lb). Tijou accepted the contract but rendered a bill nearly £1500 in excess of Talman's estimate, the latter refusing to certify the difference for payment. King William died in 1702 and Queen Anne, who did not care for Hampton Court, withdrew her patronage. Tijou removed his works from Hampton Court and continued to work at St Paul's until 1711, providing the magnificent iron screens and staircase balustrades which so enhance the cathedral. The work was, however, subject to competition and the total value extending over a period of twenty years did not exceed £4000. Tijou retired abroad leaving his wife power of attorney to collect the balance due on his work at St Paul's.

True to the fashion of his age, Jean Tijou published in 1693 *A New Book of Drawings,* illustrating his work. After his death the designs for St Paul's choir gates were adopted by the Frenchman Louis Fardrin who republished Tijou's copper plates with his own name as designer. The distinctive character of Tijou's work lies in the use of embossed acanthus leaves, rosettes, masks, garlands and crowns in great profusion. It is very unlikely that he was a practical smith, but he was certainly a very practical and artistic embosser, giving his work beauty and character never attained by any other ironworker. When his practice was at its height he must have employed many assistants, and his influence can be traced in their work. Robert Bakewell executed the delightful wrought iron garden house at Melbourne in Derbyshire, Robert Davies made the magnificent White Gates and Screen at Leeswold, near Mold in Flintshire. The work of each is perfectly distinctive and easily recognisable, but makes lavish use of acanthus and rosettes in Tijou's best manner. In internal work, however, the joiner and the carver were supreme until Robert Adam reversed the fashion and returned to the metalworker for his grilles and balustrades.

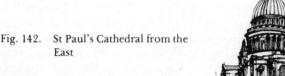

Fig. 142.   St Paul's Cathedral from the East

Joseph Rose junior was born in 1745 at Norton in Derbyshire, son, nephew and grandson of plasterers. As befitted the son of a tradesman, he was apprenticed in his father's craft and three years after he was made free of the Worshipful Company of Plaisterers in 1765, he visited Rome to study classical design at the fountainhead. He returned to England and in 1775 became Master of the Plaisterers' Company, thereafter becoming one of the best known and busiest craftsmen in England. He was influential in turning his firm's direction from the rococo decorations upon which their reputation under Rose senior had been founded in the 1740s to the new Adam style which first appeared in their work at Croome Court in Worcestershire about 1764. The Rose family built up a magnificent series of moulds in the new style and their book of friezes is now in the library of the Royal Institute of British Architects. However, at Claydon in 1768, Rose senior, while his nephew was in Italy, reverted to a version of the earlier rococo style in the saloon ceiling and cornice with medallions draped with swags of flowers. Joseph Rose and Co. monopolised English plasterwork for half a century, Rose senior organising and responsible for the administration up to his death in 1780. The firm's income was considerable. From work carried out for Robert Adam they received between the years 1761 and 1770 from work at Harewood £2829·85, in the year 1766–7 at 20 St James's Square £2684, and £1822·15 at Nostell. In addition the firm worked at Keddlestone in Derbyshire, and at Syon House in Middlesex.

When Joseph Rose junior died in February 1799, his will stipulated the disposal of his equipment and collections. These included his architectural models of vases, crests, medallions, figures, griffins, flowers and birds. Included in the sale at Christies was his collection of books, details of which make interesting reading. These books included five volumes of *Vitruvius Britannicus* by the architect Colen Campbell, Leoni's *Palladio,* and, of course, *Sketches of friezes* by Mr Joseph Rose. It is obvious that the 'plaisterer' was in the full sense of the phrase an educated man.

Glossary

ABACUS bearing plate forming the top of the capital of a classical column

ADOBE sunbaked building blocks of earth reinforced with chopped straw

AISLES areas parallel to the nave (centre aisle) of a basilica, their roofs of less height to allow for clerestorey lighting of the nave (see basilica)

ALABASTER a limestone specially valued for lending itself to fine carving, or splitting to make translucent panels

ALCOVE recess opening out of a room

AMBULATORY continuation of the aisle or aisles at the East end of mediaeval French cathedrals to form a processional path round the high altar

ANTEFIX upright blocks, ornamented on the face, to conceal the ends of tiles on the lower edge of a classical roof

APSE (APSIS) semicircular or polygonal vaulted recess at the East end of a church

AQUEDUCT artificial channel for the conveyance of water

ARCADE sequence of arches supported by columns or piers

ARCH structure spanning an opening, and holding together if only supported from the sides, transforming vertical force into lateral thrust

ASHLAR finely dressed stonework with squared edges

BASILICA (1) in Roman architecture, a public building for the administration of law

(2) in Christian architecture, a church developed from the Roman basilica with nave, aisles and apse, see plan, p. 42. Roman basilicas were used by the Jews for worship before the advent of Christianity

BATTLEMENT indented parapet of a mediaeval building, manned by archers

BILLET ornamental moulding used in Romanesque architecture, consisting of spaced rectangles or semi-cylinders

BUTTRESS structure built against a wall to give it extra support

CAME H-section strip of lead supporting glass panes in stained glass windows or leaded lights

CAPITAL the head of a column or pier

CELLA main central room in a temple (naos)

CEMENT agent binding aggregates together into concrete

CENTERING temporary structure or mould to support arched or vaulted work until it is self-supporting

CHAMFER slope across the width of a surface

CHOIR inner part of a church, usually screened off, where choristers and clergy are accommodated

CLAPBOARD weatherboarding (USA)

CLERESTOREY upper walls of the nave in a basilica, containing windows and rising above the aisle-roofs to permit direct ('clair') lighting of the nave

COB WALL wall of unburned clay mixed with chopped straw and layers of long straw as bonding, requiring protection by an overhanging roof and periodical lime wash

COFFERING recessed panelling in a ceiling

COPING weather protection for the top of a wall made of brick, concrete or stone

CORBEL cantilevered projection from the vertical face of a wall, made of stone, brick, iron, timber, etc., to carry a beam or truss

CORNICE architectural feature, projecting wholly or in part, which forms a visual and actual termination to a facade

CRAMP metal bar bent at both ends to fix or hold together masonry or timber

CRESTING carved work on the top of a building, on roof ridges and gable tops

CRUCK curved tree trunks used in rough timber frame structures

CYCLOPEAN term describing walls constructed of large irregular stone blocks

DIAPER regular decorative pattern applied to walls or panels, either painted or carved, usually conventionalised flowers in squares or lozenges

DOME convex hemispherical, circular or polygonal roof over a circular, square or polygonal plan (cupola)

ECHINUS convex moulding under the abacus (q.v.) of a Doric capital

ENTABLATURE the part of a classical temple above the columns which consists of architrave, frieze and cornice

ENTASIS the convex swelling of a column to emphasise its loadbearing function and to counteract the optical illusion that a straight column is concave

FILLET small moulding like a narrow band

FLUTES shallow rounded grooves cut perpendicularly in columns

FORMWORK temporary casing for wet concrete

FORUM the civic centre of a classical Roman town

FRESCO painting on wet lime plaster with alkali-resistant pigments dissolved in water

GILT (GILDING) covered with gold leaf

GROTTO artificial cave

HALF-TIMBERING timber frame construction with infilling of different materials or covered with weather-boarding and sometimes with lath and plaster

HIP external angle formed by the intersecting front and side slopes of a hipped roof

HYPOCAUST under-floor heating in a classical Roman building

HYPOSTYLE term describing a temple hall roof supported by a multitude of columns

IMBREX semicircular tile which covers the joint of two under-tiles

IMPOST the member below the springing line of an arch

JAMBS the vertical sides of an opening

JETTY projecting upper storey of a mediaeval timber building

KEYSTONE wedge-shaped stone at the crown of an arch, often projecting and ornamented

KILN (1) oven or furnace for the burning of brick, lime or cement

(2) heated chamber for drying timber

LANCET tall, narrow window with pointed arch, often used in groups and typical of Early English architecture

LANTERN small turret, circular, square or polygonal, open or glazed, on top of a roof or dome

LATH thin narrow slip of wood

LINE string or cord for setting out building work (plumb line)

LINENFOLD Tudor panelling with vertical grooves to resemble folded linen

LINTOL (LINTEL) horizontal member of concrete, iron or timber placed over an opening

LODGE organisation of masons engaged in mediaeval church building (as opposed to the town guilds). The lodges considered themselves religious fraternities with their own jurisdiction under the bishop or abbot

LOGGIA covered verandah, balcony or gallery

LOUVRE ventilator with sloping slats allowing passage of air and excluding rain

MASONRY stone or brick construction

MONOLITH free-standing block of stone

MORTAR mixture of lime, sand and water, or lime, cement, sand and water, used for jointing and bedding bricks, stones and slabs

MORTICE slot in a structural member to receive a projection (tenon) on one end of another member

MOSAIC floor or wall surface consisting of small pieces of marble, glass or pottery laid to a pattern or representing a picture and set in cement

MULLION vertical member dividing the lights of a window

NARTHEX vestibule forming the entrance to a basilica

NAVE main central part of a church used by the congregation

NOGGING brick panels set in timber framing

ORDER classification given by Vitruvius (q.v.) to the rules for the design of buildings in classical architecture with special regard to proportions of the Doric, Ionic and Corinthian temple elevations. An order consists of column and entablature (q.v.)

ORIEL bay window, usually supported by corbels, projecting from an outer wall in secular buildings of the late Gothic and Tudor periods

PARTY WALL now known as 'separating wall' – a wall which is common to two adjoining buildings

PEDIMENT the finishing of the gable end tops of a classical building or portico by repeating the top member of the entablature

PENDENTIVE spherical triangle solving the problem of constructing a dome over a square plan

PILE structural member of timber, r.c. or steel either constructed or driven into the ground to provide support for foundations in unsatisfactory subsoil

PINNACLE turret-like top structure of a Gothic buttress, usually ornamented and pointed

PLINTH thickening or projection at the foot of a column or wall

PORTICO open or partly enclosed area in front of the entrance to a building, with roof carried by columns, piers or posts

POST vertical structural member

POZZOLANA volcanic ash added to Roman mortar to strengthen it

PURLIN horizontal member providing intermediate support to the common rafters in roof construction

QUADRIPARTITE term applied to vaults with four separate planes

QUOIN (1) brick or stone set in a corner of a wall
    (2) external angle of a wall

RELIEVING ARCH arch built over a lintol to transfer the load to the sides of the opening and reduce the vertical pressure on the lintol

RIB stone or brick arch which carries the light, non-structural panels of a vault

RIDGE PIECE the timber member along the apex of a pitched roof which supports the top ends of the rafters

RUBBLE (1) rough undressed stones
    (2) broken bricks and similar material

RUSTICATION term applied to an elevation built of stones with rough surfaces

SPRING LINE the bedding line for the 'springer', the first stone laid in an arch

SQUINCH (-ARCH) arch built across the internal angle of a square building to support a circular or octagonal dome – the forerunner of the pendentive (q.v.)

STRAPWORK decoration consisting of interlacing bands with rivets at their intersections, fashionable in the 16th and 17th cents in Northern Europe

STRING-COURSE horizontal feature on the elevation of a building, either stone or brick, sometimes projecting

STUDS vertical timber members providing support for infilling such as wattle and daub, brick or breeze, or covered with wallboards

STYLOBATE platform for a row of columns – classical architecture

TEGULA(E) Latin word for tile, normally the name given to the undertiles of Italian roofing, their joints covered by the imbrex (q.v.)

TEMPERA mural painting employing a mixture of eggs, gum, water and pigments on a background of whiting and glue, used extensively in mediaeval times

TEMPLATE wood or metal full-size profile used to carry out some detail in plaster or other material

THOLOS circular building with a dome – classical architecture

TILE HANGING the fixing of tiles on battens to a vertical face, such as a wall or framework, to improve insulation or weather-resistance and as an architectural feature

TRABEATED term applied to architectural design based on post and beam construction (Egyptian, Greek) as opposed to arcuated architecture employing arches to span openings (Romanesque, Gothic)

TRACERY ornamental stonework in the form of geometrical patterns in the heads of Gothic windows

TRANSEPT that part of a cruciform church which lies at right angles in plan to the nave, N and S of the crossing

TRIFORIUM that part of the wide walls of the nave of a mediaeval basilica which is above the arcade and below the clerestorey lighting, often having openings to give access to the roof space of the aisles for inspection and repairs

TRIGLYPHS stone blocks with vertical grooves in the frieze of Doric temples

TRUSS rigid frame constructed of members either in tension or compression to carry a load

TUFA a porous volcanic stone found in Italy and used as an additive in Roman mortar and concrete

TYMPANUM (1) the triangular space in a pediment (q.v.)
(2) the space or panel over an opening between lintol and arch

VENEER thin layer of wood or other material applied as a decorative cover to a coarser base, or as strengthening or insulating agent

VITRUVIUS Roman architect living at the time of the Emperor Augustus and commissioned by him to write the first comprehensive manual on architecture and buildings, dealing with general principles of architecture, the development of building, the use of materials, types of buildings, civil and mechanical engineering, town planning, health, interior decoration, code of conduct and many other aspects. Its ten volumes were translated into many languages

VOLUTE spiral ornament used in Ionic, Corinthian and Composite capitals

VOUSSOIR wedge-shaped brick or stone forming part of an arch

WATTLE AND DAUB infilling to a timber framed structure, of interlaced twigs plastered with clay

ZIGGURAT Assyrian temple tower in the form of a truncated stepped pyramid

Index